Engineering Materials

This series provides topical information on innovative, structural and functional materials and composites with applications in optical, electrical, mechanical, civil, aeronautical, medical, bio- and nano-engineering. The individual volumes are complete, comprehensive monographs covering the structure, properties, manufacturing process and applications of these materials. This multidisciplinary series is devoted to professionals, students and all those interested in the latest developments in the Materials Science field, that look for a carefully selected collection of high quality review articles on their respective field of expertise.

Indexed at Compendex (2021) and Scopus (2022)

Amodini Mishra · Vinay Pathak
Editors

Smart Materials for Energy Storage and Biomedical Applications

Springer

Editors
Amodini Mishra
Indian Institute of Technology Delhi
New Delhi, India

Vinay Pathak
Indian Institute of Information Technology
Sonepat, Haryana, India

ISSN 1612-1317 ISSN 1868-1212 (electronic)
Engineering Materials
ISBN 978-3-031-92583-2 ISBN 978-3-031-92584-9 (eBook)
https://doi.org/10.1007/978-3-031-92584-9

This Springer imprint is published by the registered company Springer Nature Switzerland AG
The registered company address is: Gewerbestrasse 11, 6330 Cham, Switzerland

Dedicated to my loving parents
Mr. Jai Karan Nath Mishra *and* ***Mrs. Swayam Prabha Mishra****,*
whose unwavering love, encouragement, and support have been my greatest strength throughout this journey

Preface

In an age defined by technological advancement and innovation, the quest for materials as well as smart materials and technologies that drive electronic, photonic, magnetic, and spintronic applications has never been more critical. This book presents a comprehensive compilation of cutting-edge research, exploring the fundamental properties, synthesis techniques, and applications of various smart materials at the forefront of scientific and technological progress.

The chapters in this volume cover a broad spectrum of topics, starting with the role of magnetic doping and strong disorder in topological insulator nanostructures, shedding light on their surface state modifications. The discussion extends to recent developments in the magnetocaloric effect in pyrochlore oxides, which hold promise for energy-efficient cooling technologies.

Nanostructures, particularly plasmonic and magnetic nanoparticles, have gained tremendous attention for their potential in optoelectronic applications and oncological disease detection. The exploration of metal oxide nanostructures further highlights their significance in next-generation devices.

In the realm of optics and photonics, this book presents a theoretical analysis of optical beam defocus measurement using Zernike modes, advancements in optical fiber and photonic crystal fibers, and a deeper understanding of liquid crystal spatial light modulators for beam shaping and wavefront sensing. Moreover, the effects of hydrostatic pressure on transmittance and reflectance spectra in defective semiconductor photonic crystals are examined under TE mode.

Expanding into spectroscopy and imaging, this collection discusses photoluminescence spectroscopy—its principles, applications, and role in material characterization, as well as the relationship between optical imaging and spatial frequency. A detailed geometrical ray-tracing analysis of optical system design provides a valuable reference for researchers and engineers alike.

The final sections explore smart soft materials as a frontier in modern technology and introduce new-age magneto-dielectric and multiferroic materials, which have transformative potential in multifunctional applications.

This book aims to serve as a valuable resource for researchers, engineers, and students in materials science, optics, and nanotechnology. We extend our sincere gratitude to the contributors for their insightful chapters, which reflect the latest advancements in their respective fields. We hope that this volume inspires further research and fosters innovation in these dynamic scientific areas.

New Delhi, India Amodini Mishra
Sonepat, India Vinay Pathak

Acknowledgements

First and foremost, I am profoundly grateful to the Lord Almighty—from the depths of my soul—for his boundless mercy, grace, and unwavering guidance. His divine blessings have always illuminated my path, instilling in me the strength to persevere and uphold the virtues of righteousness.

This book is a significant milestone in my journey, and I take this opportunity to express my heartfelt appreciation to those without whom its completion would not have been possible. Words alone cannot fully convey the depth of my gratitude.

I extend my sincere and utmost gratitude to my beloved parents, whose unwavering devotion, sacrifices, encouragement, and prayers have shaped every aspect of my life. Their boundless love and support have been my greatest source of strength. No words can ever truly encapsulate the magnitude of their contributions to my journey. Endless thanks to my elder brothers and my dear sister for their continuous encouragement throughout the editing process. Their belief in me has been a driving force in this endeavor. I am also deeply thankful to my sisters-in-law for their enthusiastic moral support, which has uplifted me in countless ways. Lastly, I hold immense gratitude for my adorable nephews, whose innocent laughter and playful mischief have brightened my days and lifted the burdens of my journey. Their sweet smiles have been a source of joy and rejuvenation, reminding me of the simple yet profound beauty of family and love.

"I am deeply grateful to my lovely husband, who has been my unwavering pillar of strength throughout this journey. His constant encouragement, patience, and belief in my abilities have been instrumental in helping me navigate the challenges along the way. Through the late nights, the moments of self-doubt, and the endless hours of dedication, he stood by my side, offering his support in every possible way. His sacrifices, understanding, and reassurance gave me the confidence to pursue my passion and complete this book smoothly. Without his steadfast motivation and unconditional love, this achievement would not have been possible. I am truly fortunate to have such a loving and supportive partner, and I dedicate this milestone to him with immense gratitude and admiration"

I am also deeply thankful to the all editor of book for their assistance, cooperation, and seamless support throughout the publication process.

I sincerely acknowledge the financial support provided by the Department of Science and Technology (ANRF-DST), Government of India, under the NPDF and Women scientist (WISE-PDF) scheme. This support has been crucial in enabling my research and editorial contributions.

Additionally, this acknowledgment would be incomplete without expressing my appreciation to the entire Springer team for providing me with this wonderful opportunity and thankful to the staff of Springer publisher for their assistance, cooperation, and seamless support throughout the publication process.

Dr. Amodini Mishra

Introduction

In the ever-evolving landscape of modern science and technology, the quest for smart materials is at the forefront of innovation, driving advancements in energy storage and biomedical applications. *Smart Materials for Energy Storage and Biomedical Applications* serves as a testament to this relentless pursuit, showcasing cutting-edge research that bridges fundamental science with real-world solutions.

Each chapter in this volume offers an in depth exploration of the properties, synthesis methods, and applications of smart materials designed to revolutionize energy and healthcare sectors. From high-performance energy storage systems leveraging nanomaterials to stimuli-responsive biomaterials for targeted drug delivery and regenerative medicine, the diversity of research presented here highlights the transformative potential of these advanced materials.

As we stand on the brink of ground-breaking technological progress, the insights shared within these pages illuminate new pathways for sustainable energy solutions and next-generation biomedical innovations. From the development of self-healing batteries to advancements in bioelectronic interfaces, each contribution provides a glimpse into the forefront of scientific inquiry and technological evolution.

It is my sincere hope that the knowledge and discoveries compiled in this volume inspire researchers, engineers, and innovators to push the boundaries of possibility. May this work serve as a catalyst for interdisciplinary collaboration, driving us closer to a future where smart materials and some other important materials redefine energy efficiency, health care, and beyond.

Each chapter in this volume delves into the intricate world of materials science, exploring the fundamental properties, synthesis techniques, and applications of advanced materials designed to enhance energy efficiency and biomedical solutions. From next-generation batteries and supercapacitors to biocompatible materials for drug delivery and tissue engineering, this book highlights the transformative potential of smart materials in shaping the future.

Hydrostatic Pressure Effects on Transmittance and Reflectance Spectra in 1D Defective Semiconductor Photonic Crystals Under TE Mode Configuration 155
Chhaya Bhatia, Yogesh Sharma, Mirza Tanweer Ahmad Beig, Shikha Shukla, Rajiv Vatsa, Akanksha Namdeo, Sanjeev K. Srivastava, Amodini Mishra, and A. N. Upadhyay

Understanding Photoluminescence Spectroscopy: Principles, Applications, and Insights into Material Properties 163
Shivam Swaraj and Amodini Mishra

The Relationship Between Optical Imaging and Spatial Frequency 187
S. S. Goutam Buddha, Nagendra Kumar, Karuna S. Mallik, and Santanu Konwar

Fundamentals of Ray Tracing and Lens Design With Zemax OpticStudio 197
Karuna Sindhu Malik, Shilpa Singh, Nagendra Kumar, SSGoutam Buddha, and Santanu Konwar

Metal Oxide Nanostructures for Optoelectronic Applications 209
Sakshi Bagwari, Manish Jha, Sachin Pathak, Dheeraj K. Singh, Amodini Mishra, and Vishakha Kaushik

New Materials for New Age: Magneto-Dielectric and Multiferroic Materials 225
Anil Kumar, Anu Gupta, Rituraj Sharma, Hari Mohan Rai, M. Kamal Warshi, Vikash Mishra, Amodini Mishra, and Pankaj R. Sagdeo

About the Editors

Dr. Amodini Mishra research expertise lies in the development of advanced along with focuses on understanding the structure-property relationships towards exploring the potential applications in energy, environment, and health care.

Dr. Vinay Pathak specializes in cybersecurity, network security, and information security, integrating scientific and engineering principles to drive advanced technological innovation.

Abbreviations

1D	One Dimensional
2D	Two Dimensional
3D	Three dimensional
AFM	Atomic Force Microscopy
Al_2O_3	Aluminium oxide or Alumina
ALD	Atomic layer deposition
AR	Augmented reality
ARPES	Angle Resolved Photoemission Spectroscopy
BPM	Beam Propagation Method
BS	Beam Splitter
C	Curie constant
CBM	Conduction Band Minimum
CFRP	Carbon fibre-reinforced polymers
CGH	Computer Generated Holography
CMCE	Conventional magnetocaloric effect
CT	Computed Tomography
CTF	Contrast transfer function
CVD	Chemical vapour deposition
dB	Decibel
DCF	Dispersion-Compensating Fiber
DFT	Density Functional Theory
DM	Dzyaloshinskii-Moriya
DOEs	Diffractive optics elements
DOS	Density of States
DSF	Dispersion-Shifted Fiber
DSPhC	Defective Semiconductor Photonic crystal
DWDM	Dense Wavelength-Division Multiplexing
EAF	Electrostatic force
EAPs	Electroactive polymers
EDFA	Erbium-Doped Fiber Amplifier
EM	Electromagnetic

ZnO	Zinc Oxide
ΔSM	Change in magnetic entropy
θ_{CW}	Curie-Weiss temperature

Role of Magnetic Doping and Strong Disorder on Surface States of Topological Insulator Nanostructures

Anu Gupta, Anil Kumar, Shivam Shukla, Rituraj Sharma, Amodini Mishra, and Sanjeev Kumar Srivastava

Abstract Topological Insulators (TIs) are a type of quantum material which have conducting surface states and insulating bulk states. Time-reversal symmetry (TRS) protects the gapless surface states from disorder and non-magnetic impurity backscattering. On the other hand, magnetic impurities like disorders such as anti-site and vacancy defects disrupt the TRS and result in intriguing phenomena like the quantum anomalous Hall effect, magnetoelectric effect, and magnetic monopole. The advantage of studying TIs in nanostructures is that their huge surface-to-volume ratio amplifies the features that are dominated by the surface. Bi_2Se_3, one of the most well-known TIs, has been the focus of a great deal of research due to its energy gap of 0.3 eV, which is greater than the room temperature energy scale. The main goal of this book chapter is to examine how strong disorder and magnetic doping affect the surface state of Bi_2Se_3 nanostructures. The exact magnetic and transport features in this system, which are backed by theoretical calculations, will be covered in depth.

Keywords Topological insulators · Magnetic doping · Disorder · Nanostructures

Anu Gupta, Anil Kumar, Shivam Shukla, these authors have contributed equally

A. Gupta (✉)
Department of Chemical and Biological Physics, Weizmann Institute of Science, Rehovot, Israel
e-mail: anu.gupta@weizmann.ac.il

A. Gupta · S. Shukla · S. K. Srivastava
Department of Physics, Indian Institute of Technology Kharagpur, Kharagpur, India

A. Kumar
Department of Physics, Indian Institute of Technology Kanpur, Kanpur, India

R. Sharma
Centre for Scientific and Applied Research (CSAR), IPS Academy, Indore, India

A. Mishra
Department of Physics, Indian Institute of Technology Delhi, New Delhi, India

A. Mishra and V. Pathak (eds.), *Smart Materials for Energy Storage and Biomedical Applications*, Engineering Materials, https://doi.org/10.1007/978-3-031-92584-9_1

1 Introduction

A few years ago, the discovery of Topological Insulators (TIs) lead to advanced research in condensed matter physics, gathered a large number of theorists and experimentalists at once. TIs are an exotic state of quantum state that differs from ordinary insulators in that electrons cannot conduct in the bulk of the material while at surface or edge states, they do conduct. Basically, Time-reversal symmetry (TRS) limits the metallic or conducting electronic states that are supported by the surfaces of three-dimensional (3D) TIs and the edges of two-dimensional (2D) TIs [1–6]. These conducting states exhibit special features i.e. chirality, or locking of the spin and momentum directions, which do not allow the backscattering or U-turns when there are no magnetic impurities present. Figure 1 explains the propagation of electrons in 2D and 3D TIs. The helicity or chirality of surface states originates from spin–orbit coupling (SOC) which occurs in heavy elements such as bismuth, tellurium, or antimony. The band structure has a band inversion as a result of significant SOC. However, the topological TRS protection is broken when magnetic contaminants and strong disorder are present in the system which leads to variety of intriguing phenomena, including the topological magneto-electric effect [7, 8], magnetic monopole [9] and the quantum anomalous Hall effect (QAHE) [10]. Bi_2Se_3 is one of most widely studied 3D TIs with an energy gap of 0.3 eV, which is topologically non-trivial and greater than the room temperature energy scale [11].

TIs with magnetic doping and strong disorder should exhibit a wide range of fascinating behaviors. TIs with magnetic impurities exhibit a range of magnetic phases, including spin glass, ferromagnetic, antiferromagnetic, and paramagnetic. Different kinds of exchange interactions between the magnetic impurities in some surface and bulk states give birth to these phases. Transition metals and lanthanides, or rare-earth elements, are the primary magnetic elements. The elements that are most frequently employed for magnetic doping in transition metals are Fe, Mn, Cr, Co, Ni, and V. Especially in the case of ferromagnetism, ferromagnetic ground state breaks TRS when directed perpendicular to the TI surface and opens a gap in the surface state as shown in Fig. 2. Defects such as vacancies, antisites, and nonstoichiometric doping are the main causes of disorder in TIs [12, 13]. Localized states are created when defects are added to TIs, and the electrons in these states act differently from the conduction electrons. Additionally, it is reported that the TI loses its capacity to shield surface states and transition into trivial insulating states above a certain threshold of disorder [14]. However, annealing controls the reverse conversion process that returns material to its initial state.

For a wide range of industrial applications and for exploring the underlying nature of the TI surface states, TI nanostructures provide several benefits over their bulk counterparts. Because of the high surface-to-volume ratio, improved surface effects are a nanomaterial's most evident benefit. TIs have been fabricated into a variety of nanostructures by various groups, including exfoliated flakes from bulk crystals [15], nanoribbons [16], and nanoplates [13, 17, 18]. TI nanostructures' enhanced surfaces offer an excellent platform for researching the characteristics of surface electron

The TEM image for the sample is displayed in Fig. 4a. The TEM image shows the hexagonal shape of the nanoplates. The lateral dimensions of the hexagonal shaped nanoplates have been determined to be between 100 and 150 nm. Figure 4b shows the M–H data which exhibits two components. One component is diamagnetic, indicated by the negative slope, which is expected for Bi_2Se_3. The other component is paramagnetic and follows a Brillouin-function related behavior, as displayed in the inset of Fig. 4b. Bi-rich Bi_2Se_3 nanoplates have also shown similar M–H curves, as reported in previous studies [32]. This phenomenon has been linked to an intriguing orbital magnetism. Due to the noisy data, it is evident that the observed hysteresis loop is solely a result of fluctuated data points and cannot represent any potential magnetic ordering in the sample. The observed saturation magnetization of 1.37×10^2 emu/g is corresponding to a magnetic moment of ~ 1.6×10^3 μB/f.u. It is crucial to determine the source of these moments. Regarding this matter, it is worth noting that the defects commonly found in Bi_2Se_3, such as Se vacancies and Bi_{Se} anti-sites, can be identified through DFT computations as potential sources of moments [13, 29]. The DOS calculations of Bi vacancies and anti-site defects, as manifested in Fig. 4c and d, exhibit asymmetry, indicating the magnetic nature of these defects. Based on the calculations, a magnetic moment of 1.0 μB can be assigned to each Bi vacancy, while each Bi_{Se} anti-site defect also has a magnetic moment of 1.0 μB. The presence of these moments provides a qualitative explanation for the observed paramagnetism in the Bi_2Se_3 nanoplates. For a more precise analysis, let's assume that there are roughly equal amounts of Bi_{Se} anti-site and Bi vacancies defects in the nanostructures. The calculated value of magnetic moment of per defect is 1.0 μB. The measured moment of 1.6×10^3 μB/f.u. can be explained by a defect density of 10^{19} defects/cm^3, which is identical to 0.16 atomic percent. The defect density lies within the range of typical semiconductor dopant densities, which is quite high [34]. It shows that the pure Bi_2Se_3 nanoplates have a high defect density, which indicates that the system is highly disordered. Given the significant defect density, the electron scattering from these defects becomes quite noticeable. As a result, the resistivity increases significantly, making it difficult to observe any surface conduction on the metallic surface, despite the large surface-to-volume ratio in the nano dimension sample.

The doped samples exhibit a clear paramagnetic nature, as can be seen from the Magnetization versus Field curves depicted in Fig. 5a and b. The curves of each composition exhibit a transition from linear paramagnetic behavior to Brillouin-function type paramagnetic behavior as the temperature reduces from high to low. At low temperatures, the magnetizations exhibit a tendency to reach saturation levels beyond 4 T. The saturation moment values for $x = 0.1$ and 0.3 are ~ 1.3 μB/f.u. and ~ 2.2 μB/f.u., respectively.

It would be beneficial to have spin-polarized DOS calculations for $x = 0.1$ and 0.3 to consider these moments. Figure 4c and d display the spin-polarized total and Dy f DOSs, respectively. It is evident that the DOS of both spins in both cases exhibit asymmetry, primarily due to the presence of the Dy *f localized* states. It further confirms that doped Dy atoms have magnetic moments.

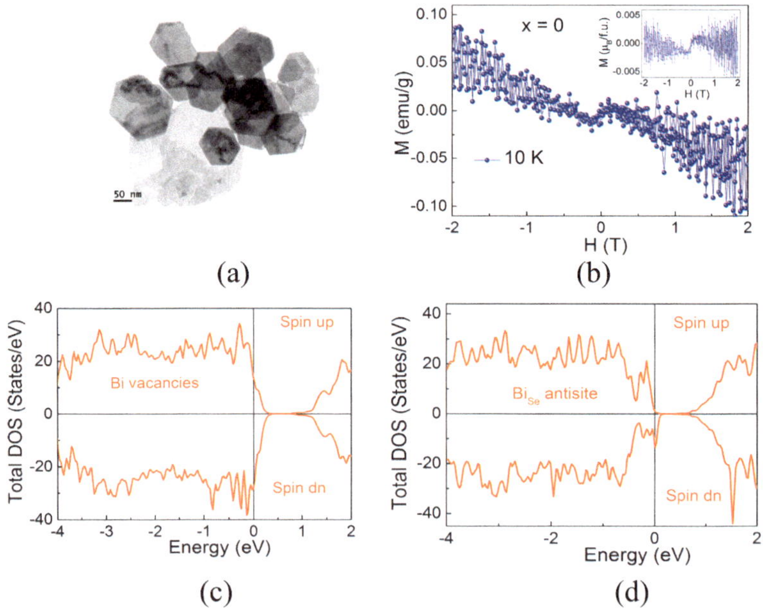

Fig. 4 **a** The TEM micrograph illustrates the $x = 0.3$ nanoplates. **b** The M–H plot of Bi_2Se_3 is shown, the inset displays the M–H after subtracting the diamagnetic component, **c** Total Spin-polarized DOS of Bi vacancies defected Bi_2Se_3. **d** Total Spin-polarized DOS of Bi_{Se} anti-site defected Bi_2Se_3. Reproduced from [31], with the authorization of AIP Publishing

The computed Dy moment in $x = 0.083$ and 0.33 structures are approximately ~ 4.7 μB. Given that 0.16% of defects in Bi_2Se_3 each have a moment of 1.0 μB, it results in an observed moment of 1.6×10^3 μB/f.u. If we apply a rough linear scaling, we can expect magnetic moment values of 0.5 and 1.5 μB/f.u. in samples with x values of 0.1 and 0.3, respectively. Therefore, the observed values (1.33 and 2.25 μB/f.u.) are close to the estimated ones.

4.2 Transport Properties and Role of Annealing

The localization characteristics of undoped and doped samples are further investigated through the magneconductance measurements. The magnetoconductance is represented by Eq. (1)

$$\Delta G(H) = G(H) - G(0) \tag{1}$$

where $G(H)$ and $G(0)$ are the conductance at field H and zero-field. The applied magnetic field direction is perpendicular to the applied electric current. Figure 6a

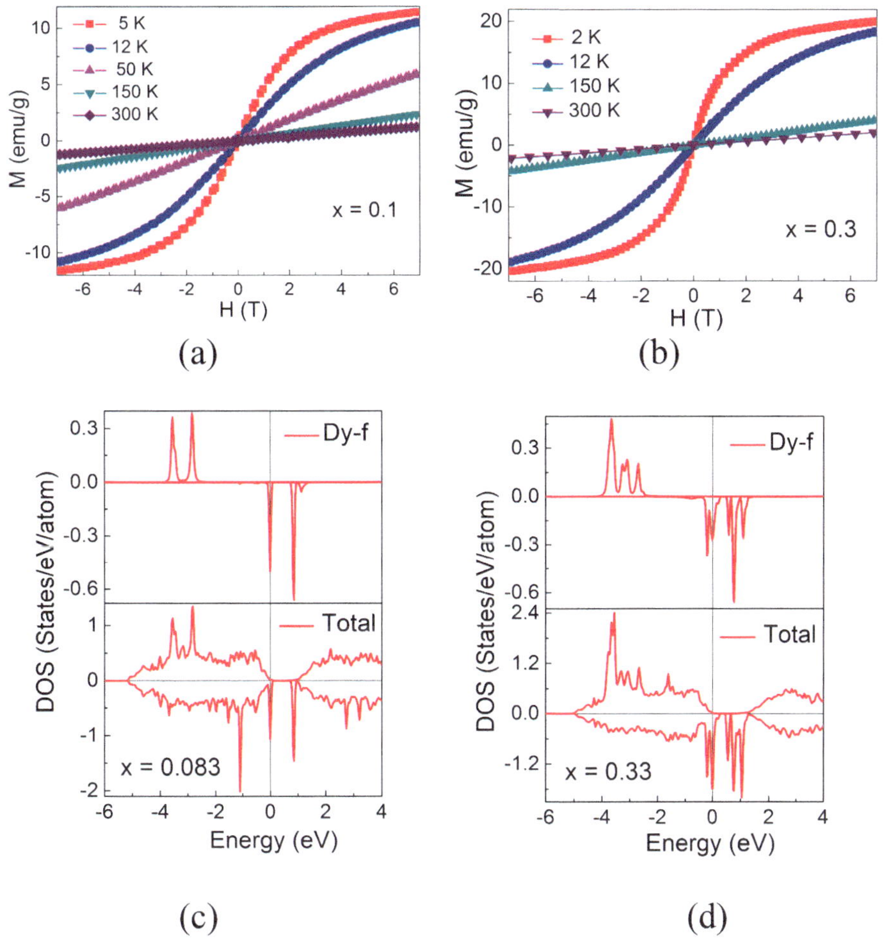

Fig. 5 Magnetizations versus Field curves at different temperature for **a** x value of 0.1. **b** x value of 0.3. DOSs for **c** $x = 0.083$ and **d** $x = 0.33$. Reprinted from [31], with the permission of AIP Publishing

shows the magnetoconductance of all the different concentrations of doping at a temperature of 20 K. The positive magnetoconductance signifies the WL nature of nanoplates. To further confirm WL and finding the additional parameters in it, the MC is fitted with the Hikami-Larkin-Nagaoka (HLN) equation [38].

$$\Delta G(H) = \frac{\alpha e^2}{\pi h}\left[\Psi\left(\frac{\hbar}{4eHl_{\emptyset}^2} + \frac{1}{2}\right) - \ln\left(\frac{\hbar}{4eHl_{\emptyset}^2}\right)\right] \tag{2}$$

Here ψ is the digamma function, l_{Φ} is the phase coherence length, and α is a prefactor that provides information about the nature of electrons. The fitting from Eq. (2) is

10. Yu, R., Zhang, W., Zhang, H.-J., Zhang, S.-C., Dai, X., Fang, Z.: Quantized anomalous hall effect in magnetic topological insulators. Science **329**, 61–64 (2010)
11. Xia, Y., Qian, D., Hsieh, D., Wray, L., Pal, A., Lin, H., Bansil, A., Grauer, D., Hor, Y.S., Cava, R.J., Hasan, M.Z.: Observation of a large-gap topological- insulator class with a single Dirac cone on the surface. Nat. Phys. **5**, 398–402 (2009)
12. West, D., Sun, Y., Wang, H., Bang, J., Zhang, S.: Native defects in second generation topological insulators: effect of spin-orbit interaction on Bi_2Se_3. Phys. Rev. B **86**, 121201 (2012)
13. Gupta, A., Chowdhury, R.K., Ray, S.K., Srivastava, S.K.: Selective photoresponse of plasmonic silver nanoparticle decorated Bi_2Se_3 nanosheets. Nanotechnology **30**, 435204 (2019)
14. Brahlek, M., Koirala, N., Salehi, M., Moon, J., Zhang, W., Li, H., Zhou, X., Han, M.G., Wu, L., Emge, T., Lee, H.D.: Disorder-driven topological phase transition in Bi_2Se_3 films. Phys. Rev. B **94**(16), 165104 (2016)
15. Teweldebrhan, D., Goyal, V., Balandin, A.A.: Exfoliation and characterization of bismuth telluride atomic quintuples and quasi-two-dimensional crystals. Nano Lett. **10**, 1209–1218 (2010)
16. Kong, D., Randel, J.C., Peng, H., Cha, J.J., Meister, S., Lai, K., Chen, Y.L., Shen, Z.X., Manoharan, H.C., Cui, Y.: Topological Insulator Nanowires and Nanoribbons. Nano Lett. **10**, 329–333 (2010)
17. Kong, D., Dang, W., Cha, J.J., Li, H., Meister, S., Peng, H., Liu, Z., Cui, Y.: Few-layer nanoplates of Bi_2Se_3 and Bi_2Te_3 with highly tunable chemical potential. Nano Lett. **10**, 2245–2250 (2010)
18. Gupta, A., Hillier, A.D., Telling, M.T.F., Srivastava, S.K.: Local magnetic behaviour of highly disordered undoped and Co-doped Bi_2Se_3 nanoplates: a muon spin relaxation study. Nanotechnology **33**, 215701 (2022)
19. Bergmann, G.: Weak localization in thin films: a time-of-flight experiment with conduction electrons. Phys. Rep. **107**, 1–58 (1984)
20. Moore, J.E., Balents, L.: Topological invariants of time-reversal-invariant band structures. Phys. Rev. B **75**, 121306 (2007)
21. Hsieh, D., Qian, D., Wray, L., Xia, Y., Hor, Y.S., Cava, R.J., Hasan, M.Z.: A topological Dirac insulator in a quantum spin hall phase. Nature **452**, 970 (2008)
22. Fu, L., Kane, C.L.: Topological insulators with inversion symmetry. Phys. Rev. B **76**, 045302 (2007)
23. Anderson, P.W.: Absence of diffusion in certain random lattices. Phys. Rev. **109**, 1492 (1958)
24. Abrahams, E., Anderson, P.W., Licciardello, D.C., Ramakrishnan, T.V.: Scaling theory of localization: Absence of quantum diffusion in two dimensions. Phys. Rev. Lett. **42**, 673 (1979)
25. Lee, P.A., Ramakrishnan, T.V.: Disordered electronic systems. Rev. Mod. Phys. **57**, 287–337 (1985)
26. Zhang, T., Cheng, P., Chen, X., Jia, J.-F., Ma, X., He, K., Wang, L., Zhang, H., Dai, X., Fang, Z., et al.: Experimental demonstration of topological surface states protected by time-reversal symmetry. Phys. Rev. Lett. **103**, 266803 (2009)
27. Roushan, P., Seo, J., Parker, C.V., Hor, Y.S., Hsieh, D., Qian, D., Richardella, A., Hasan, M.Z., Cava, R.J., Yazdani, A.: Topological surface states protected from backscattering by chiral spin texture. Nature **460**, 1106 (2009)
28. Fleurial, J.P., Gailliard, L., Triboulet, R., Scherrer, H., Scherrer, S.: Thermal properties of high quality single crystals of bismuth telluride-Part I: experimental characterization. J. Phys. Chem. Solids **49**, 1237–1247 (1988)
29. Xue, L., Zhou, P., Zhang, C.X., He, C.Y., Hao, G.L., Sun, L.Z., Zhong, J.X.: First-principles study of native point defects in Bi_2Se_3. AIP Adv. **3**, 052105 (2013)
30. Blaha, P., Schwarz, K., Madsen, G.K., Kvasnicka, D., Luitz, J.: Technische Universität, (2001)
31. Gupta, A., Srivastava, S.K.: Paramagnetism, hopping conduction, and weak localization in highly disordered pure and Dy-doped Bi_2Se_3 nanoplates. J. Appl. Phys. **127**, 244302 (2020)
32. Kim, H.J., Katsiotis, M.S., Alhassan, S., Zafiropoulou, I., Pissas, M., Sanakis, Y., Mitrikas, G., Panopoulos, N., Boukos, N., Tzitzios, V., et al.: Unexpected orbital magnetism in Bi-rich Bi_2Se_3 nanoplatelet. NPG Asia Mater. **8**, e271 (2016)

33. Scanlon, D., King, P., Singh, R., De La Torre, A., Walker, S.M., Balakrishnan, G., Baumberger, F., Catlow, C.: Controlling bulk conductivity in topological insulators: key role of anti-site defects. Adv. Mater. **24**, 2154–2158 (2012)
34. Li, S.S., Thurber, W.R.: The dopant density and temperature dependence of electron mobility and resistivity in *n*-type silicon. Solid State Electron. **20**, 609–616 (1977)
35. He, H.-T., Wang, G., Zhang, T., Sou, I.-K., Wong, G.K.L., Wang, J.-N., Lu, H.-Z., Shen, S.-Q., Zhang, F.-C.: Impurity effect on weak antilocalization in the topological insulator Bi_2Te_3. Phys. Rev. Lett. **106**, 166805 (2011)
36. Zheng, G., Wang, N., Yang, J., Wang, W., Du, H., Ning, W., Yang, Z., Lu, H.-Z., Zhang, Y., Tian, M.: Weak localization effect in topological insulator micro flakes grown on insulating ferrimagnet $BaFe_{12}O_{19}$. Sci. Rep. **6**, 21334 (2016)
37. Liu, M., Zhang, J., Chang, C.-Z., Zhang, Z., Feng, X., Li, K., He, K., Wang, L.-L., Chen, X., Dai, X., Fang, Z., Xue, Q.-K., Ma, X., Wang, Y.: Crossover between weak antilocalization and weak localization in a magnetically doped topological insulator. Phys. Rev. Lett. **108**, 036805 (2012)
38. Tung, Y., Chong, C., Liao, C., Chang, C., Huang, S., Chuang, P., Lee, M., Cheng, C., Li, Y., Liu, C.-P., et al.: Tuning the transport and magnetism in a Cr–Bi_2Se_3 topological insulator by Sb doping. RSC Adv. **7**, 47789 (2017)
39. Wang, H., Liu, H., Chang, C.-Z., Zuo, H., Zhao, Y., Sun, Y., Xia, Z., He, K., Ma, X., Xie, X., et al.: Crossover between weak antilocalization and weak localization of bulk states in ultrathin Bi_2Se_3 Films. Sci. Rep. **4**, 5817 (2014)
40. Zhang, L., Dolev, M., Yang, Q.I., Hammond, R.H., Zhou, B., Palevski, A., Chen, Y., Kapitulnik, A.: Weak localization effects as evidence for bulk quantization in Bi_2Se_3 thin films. Phys. Rev. B, **88**, 121103 (2013)

Recent Development of Magnetocaloric Effect in Pyrochlore Oxides

Vinod Kumar Dwivedi

Abstract In recent years, the conventional gas compression-expansion based refrigeration technology utilizes hazardous gases. Therefore, there is urgent requirement of alternative potential candidate materials for cooling technology. One of the proposed promising alternative materials to the conventional cooling technique is solid state magnetic cooling based on magnetocaloric effect (MCE). In this chapter, the current developments of MCE in pyrochlore oxide materials are presented. Pyrochlore oxides exhibit various complex magnetic states gives a broad template to explore these materials in MCE point of view. Pyrochlore oxides show strong frustration due to its crystal structure. All reported litterateurs about MCE in pyrochlore oxides are $A_2Mn_2O_7$ (A = Er, Dy, Ho, Yb, La, Fe, Al, Ga, Sn), $Er_2Ti_2O_7$, A_2TiMnO_7 (A = Er, Dy, Ho), $A_2Mo_2O_7$ (A = Er, Dy, Gd, Y), $Pr_2Sn_2O_7$, $Er_2Ir_2O_7$ and $Y_2Ir_{2-x}Cr_xO_7$. Among all compounds, cubic pyrochlore $Er_2Mn_2O_7$ exhibit largest conventional MCE value at change of field 5 T. On the other hand, $Y_2Ir_{2-x}Cr_xO_7$ series show coexistence of inverse and conventional magnetocaloric effect which is absent in whole reported family of pyrochlore oxides. It is worth mention that depending on synthesis condition $A_2Mn_2O_7$ series crystallizes in either cubic pyrochlore phase ($Fd\bar{3}m$) or monoclinic layered perovskite (P2/M). It suggests the need of more materials to be explore experimentally and theoretically to establish the connection between these two phases.

Keywords Magnetocaloric effect · Pyrochlore oxides · Frustration · Pyrochlore iridates

1 Introduction

Now a days, in residential and commercial areas refrigeration technology has largest energy consumption rate i.e., 31% (USA), 40% (Europe) and 17% (across the globe) [1]. The demand for refrigeration technology is expected to substantially grow with

V. K. Dwivedi (✉)
Department of Materials Science and Engineering, Tel Aviv University, Tel Aviv, Israel
e-mail: vinodd.iitbombay@gmail.com

A. Mishra and V. Pathak (eds.), *Smart Materials for Energy Storage and Biomedical Applications*, Engineering Materials, https://doi.org/10.1007/978-3-031-92584-9_2

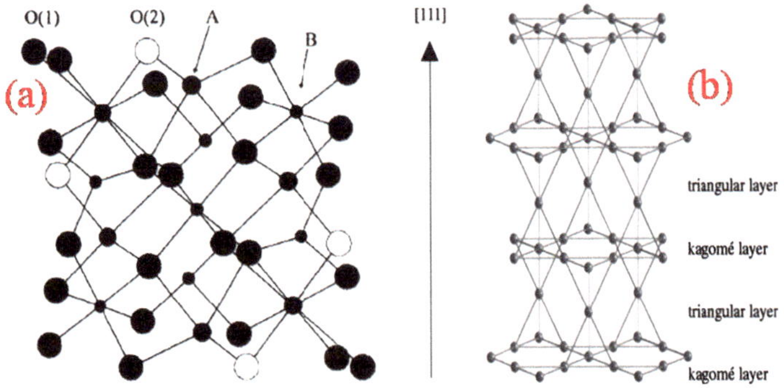

Fig. 1 **a** Structure of $A_2B_2O1_6O2$ pyrochlore. **b** The view of [111] direction stacked alternating kagome and triangular layers. All pyrochlore iridates show trigonal com- pression of the oxygen cages along the [111] directions. Adapted from [28, 29]

the lattice stability significantly. Therefore, for oxygen deficient pyrochlores, partial loss of the O2- anions is supported [28, 29].

2.3 *Magnetic Frustration*

In condensed matter physics, either frustration or competing interactions are common. Frustration arises when a system is unable to minimize its total classical energy [29–31]. Figure 2a shows spins reside on the square lattice easily satisfy the requirement by minimizing its total classical energy leading to absence of frustration. Frustrated magnetic systems can be classified in two types. The first type is because of the lattice geometry and the second in which site or bond randomness is a key factor. Figure 2b shows simplest case of geometric frustration. Ising spins are placed at the vertices of an equilateral triangle pointing either up or down direction, such spins interact via nearest neighbour antiferromagnetic exchange interaction. The tetrahedron shown in Fig. 2c is three-dimensional version of a scenario similar to triangular lattice with four nearest neighbour spins situated at tetrahedron vertices. The triangular lattice made of many edge-sharing triangles, has very large degeneracy or configurational spin disorder, which leads not only to a zero-temperature residual entropy but absence of symmetry breaking phase transition down to absolute zero temperature. This is an example of geometric frustration where the regular periodic configuration (triangular, kagome, tetrahedral or otherwise) of the space lattice prevent to develop a long range ordered state. The same system containing uniaxial Ising spins with ferromagnetic interaction does not show frustration.

The second type of frustration comes into picture due to ‘quenched randomness’. The random- ness in systems can be introduced by many ways. Other than impurities by doping, emergence of vacancies, grain boundaries or dislocations in a crystal give rise to disorder in a system. Besides reducing long range ordering temperature, the

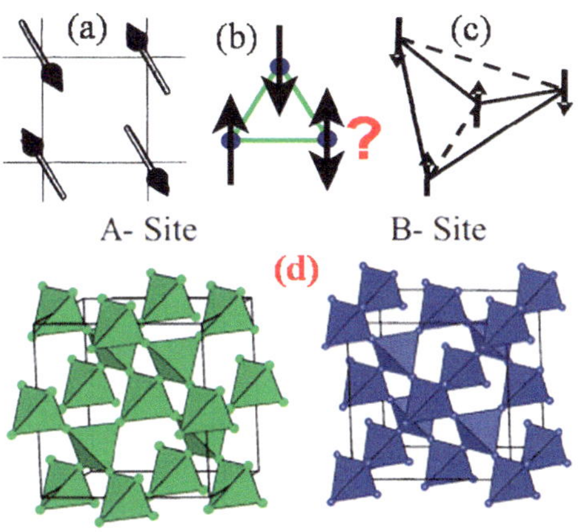

Fig. 2 Illustration of antiferromagnetic nearest neighbour interactions at **a** square lattice, **b** triangular lattice and **c** tetrahedron. Ising spin on a tetrahedron exhibit four satisfied antiferromagnetic and two unsatisfied ferromagnetic frustrated bonds (dashed line). **d** Structure of A and B sublattices of cubic pyrochlore, where either both or one sublattices can be magnetic. Adapted from [29, 30]

quenched or frozen-in disorder systems where defects or impurity atoms do not move over considerable time scales could also lead to non-trivial magnetic phases such as spin glass, Griffiths phase [31, 32], etc. The dilution of magnetic moments (or substitution of non-magnetic atoms) in a disorder free geometrically frustrated system can also generate random frustration. A spin glass system can be defined as a magnetic system with randomly mixed-interaction. Such system shows a transition from thermally fluctuating spins of paramagnetic state to a glass-like state (where spins get frozen in time but irregular in direction below a certain temperature called freezing temperature) [31]. In contrast to pure magnets (disorder free), spin-glass does not show long range magnetic ordering. The most common examples of spin glass systems are $Cu_{1-x}Mn_x$ and AuFe, in which a very little amount of magnetic impurity ($x \sim 1\%$) is introduced in non-magnetic host Cu and Au, respectively [31].

Experimentally, magnetic frustration strength (f) in magnetic system can be estimated as [30]

$$f = \frac{|\theta_{CW}|}{T^*} \tag{1}$$

where fitting of high temperature paramagnetic part of inverse dc magnetic susceptibility using following Eq. 2 gives Curie–Weiss temperature θ_{CW}.

$$\chi = \frac{C}{T - T^*} \tag{2}$$

For spin glass, T^* would be the freezing temperature T_f for spin glass system, C is Curie constant. Heavily frustrated system has very low T^* as compared to θ_{CW}.

The geometrical frustration is closely connected to the pyrochlore oxides. In the cubic pyrochlore oxides, $A_2B_2O_7$ (A = rare earth trivalent element, Y, and sometimes Sc Y, Bi, and B = transition/p-block metals), A and B ions lie on two distinct interpenetrating lattices made up with corner-sharing tetrahedra (Fig. 2d). In $A_2B_2O_7$, if A, B, or both ions are magnetic and exchange interaction between nearest neighbour is antiferromagnetic, then the system can be highly frustrated by geometry. Therefore, pyrochlore oxides exhibiting antiferromagnetically coupled spins, may not develop a long-range magnetic ordering. Examples of such phenomena exhibited by pyrochlore oxides are spin glass in $Y_2Mo_2O_7$ [33], spin liquid behaviour in $Tb_2Ti_2O_7$ [34], spin-ice phenomena [35] in $Ho_2Ti_2O_7$, $Dy_2Ti_2O_7$ and $Tb_2Sn_2O_7$ [36], unconventional anomalous Hall effect in metallic $Nd_2Mo_2O_7$ [37, 38], superconductivity in $Cd_2Re_2O_7$ [39] and the possible frustrated Kondo-lattice like effect in $Pr_2Ir_2O_7$ [40, 41], etc. These few examples show the importance of geometrical frustration in pyrochlore oxides.

3 Estimation of Important MCE Parameters

The process of calculation for the few important parameters are given below.

(I) The change in magnetic entropy ΔS_M can be calculated using Maxwell relation [7] from the M–H isotherms

$$\Delta S_M = \int_0^H (\partial M/\partial T)dH \tag{3}$$

(II) The potential for magnetic cooling generally referred as relative cooling power (RCP) can be estimated using following relation.

$$\mathrm{RCP} = —\Delta S_M \max \delta T_{\mathrm{FWHM}} \tag{4}$$

(III) The capacity of refrigerant (RC) defined as the amount of heat transferred between the hot end T_{hot} and cold end T_{cold}. And can be calculated by estimating the area enclosed by $-\Delta S_M$ versus T curve between T_{hot} and T_{cold} using Eq. 5 for a particular magnetic field as shown in Fig. 3.

$$RC = \int_{cold}^{hot} [\Delta S_M(T)]dT \tag{5}$$

The two temperatures T_{cold} and T_{hot} (i.e., full width at half maximum δT_{FWHM} of $-\Delta S_M$ versus T curve) are associated to the working window of the refrigerator. The materials exhibiting larger RC is preferred to transfer the larger amount of heat as compared with distinct materials employed in same thermodynamic cycle.

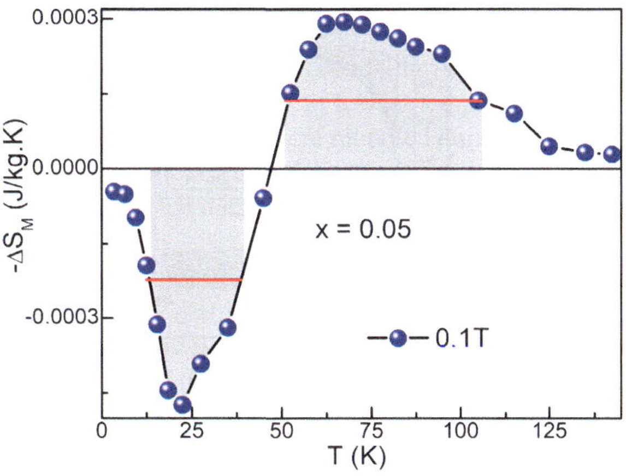

Fig. 3 Calculation method for RCP and RC from ΔS_M versus T curve

4 MCE in Pyrochlore Oxides $A_2B_2O_7$

Long back, theoretically it has been predicted that the classical Heisenberg antiferromagnetic system exhibiting strong magnetic frustration could host large adiabatic temperature change induced by applied field as compared to non-frustrated magnetic systems [26, 27]. Pyrochlores are known to exhibit strong magnetic frustration. It has drawn the attention of researchers worldwide to investigate the MCE in such highly frustrated pyrochlore systems. The following section discuss the progress of MCE in various pyrochlore oxide systems.

4.1 Mn Based Pyrochlore Oxides

The first MCE investigation was done on the compound $Er_2Mn_2O_7$ [42]. In this material, 4f- localized magnetic moments of Er^{3+} and 3d-metal electrons of Mn^{4+} both are magnetic giving rise to a competition between the magnetic interactions. Synthesis of polycrystalline $Er_2Mn_2O_7$ have performed by solid-state reaction method heated at 1300 °C for 48 h in air using the ingredients Er_2O_3 and MnO_2 reported elsewhere [42]. The refinement of X-ray diffraction (XRD) pattern suggests monoclinic phase with space group P2/M. Figure 4a shows enhancement in magnetization (M) versus temperature (T) curve measured at various magnetic field, suggest the presence of ferromagnetic components. The antiferromagnetic transition temperature known as Neel temperature T_N can be determined from minimum in first derivative of magnetization (dM/dT versus T curve) demonstrated in inset of Fig. 4a.

The values of T_N are 1.93 K (0 T), 1.97 K (0.05 T), 2.45 K (1 T), 3.25 K (2 T), 4.25 K (3 T), 5.01 K (4 T) and 5.47 K (5 T).The estimated values of θ_{CW} are – 24.05 K (0.05 T), – 16.55 K (1 T), – 15.16 K (2 T), – 14.47 K (3 T), – 14.23 K (4 T) and – 13.92 K (5 T). The negative values of θ_{CW} indicates antiferromagnetic (AFM)

16.1 (5 T), 18.0 (6 T), and 20.1 (7 T) J/K-kg shown in Fig. 5d. It can also be seen that RC (RCP) versus H increases linearly (Fig. 5d) and reaches to the value 68 (90), 288 (377), 522 (680), and 747 J/kg (964 J/kg) under change of applied field 1, 3, 5, and 7 T, respectively. The value of RC is quite large for $Er_2Mn_2O_7$ compound. Such a giant RC is associated to the broad ΔS_M-T curves imply the wide working temperature window ranging 20–80 K. In addition, $Er_2Mn_2O_7$ possess a large value of $\Delta S_M = 5.3$ J/K-kg and RC = 68 J/kg at low field change (~ 1 T), makes it a potential low field MCE candidate suitable for the permanent magnets.

In pyrochlore $Er_2Mn_2O_7$, Mn^{4+} ($3d^3$, S = 3/2) sublattices order ferromagnetically due to Mn-O-Mn bond angle 130° (favors FM alignment). It reduces the strength of frustration, yields strong internal field to align Er^{3+} ($4f^{11}$, S = 3/2, L = 6, J = 15/2) sublattices. It leads to a two-sublattice cooperative FM alignment of spins giving rise to large reversible MCE at T_C. It is obvious that two sublattice cooperative FM ordering plays an important role for the giant MCE values in cubic pyrochlore oxide $Er_2Mn_2O_7$ with space group Fd¯3m synthesized at high pressure high temperature. While it is absent in the isomeric $Er_2Mn_2O_7$ compound crystallizes in a monoclinic structure with space group P2/M prepared at ambient pressure [42]. Pyrochlore $Er_2Mn_2O_7$ exhibit FM transition at $T_C = 34.5$ K, while monoclinic layered perovskite $Er_2Mn_2O_7$ goes under AFM transition at $T_N = 2$ K.

Further, the MCE of other members of FM pyrochlore oxides $A_2Mn_2O_7$ (A = Dy, Ho, Yb) series have also been investigated [46]. Here, Dy^{3+}, Ho^{3+}, and Yb^{3+} have larger magnetic moment than Er^{3+}. The $A_2Mn_2O_7$ cubic pyrochlore series can be prepared under high-pressure high temperature conditions. Whole series show a cooperative SOMPT FM transition at $T_C \approx 39$ K (Fig. 6). Similar to $Er_2Mn_2O_7$ cubic pyrochlore oxide, $A_2Mn_2O_7$ (R = Dy, Ho, Yb) series display reversible and large value of MCE around T_C. The maximum value of ΔS_M (RC) under 7 T magnetic field change for $A_2Mn_2O_7$ series reaches to 20.1 J/K-kg (564 J/kg) for $Dy_2Mn_2O_7$, 18.4 J/K-kg (503.7 J/kg) for $Ho_2Mn_2O_7$ and 13.9 J/K-kg (371.4 J/kg) for $Yb_2Mn_2O_7$. The series possess lesser values than expectation (larger magnetic moments of Dy^{3+}, Ho^{3+}, and Yb^{3+} ion compared to Er^{3+}). In pyrochlore $Er_2Mn_2O_7$, MCE phenomena associated to two-sublattice cooperative FM ordering, while for $A_2Mn_2O_7$ (R = Dy, Ho, Yb) series, strong single-ion anisotropy prevents the complete FM ordering of magnetic moments of A^{3+} ion.

In a recent study, the effect of ionic radius on the evolution of MCE and Griffiths phase have been performed in new type of compound $A_2Mn_2O_7$ (A = La, Fe, Al) [47–50]. The series of materials were prepared at ambient pressure. The refinement of XRD pattern suggests monoclinic layered perovskite structure with P2/M space group. All three samples $La_2Mn_2O_7$, $Fe_2Mn_2O_7$, and $Al_2Mn_2O_7$ show a FM transition and Griffiths-like state at low temperature (Fig. 7). Figure 7d display ΔS_M-T curves for 5 T filed change. It is obvious that the value of ΔS_M, RCP and strength of Griffiths-like phase enhances by reducing the ionic radius of A^{3+}-site.

Very recently the magnetic and MCE properties have been explored in $Ga_2Mn_{2-x}Cr_xO_7$ ($x = 0.0$, 0.1, 0.3, and 0.5) series compound [51]. Samples were prepared at ambient pressure and found to be crystallize in cubic pyrochlore phase exhibiting space group Fd¯3m. All samples display AFM transition at T_N ~ 42 K

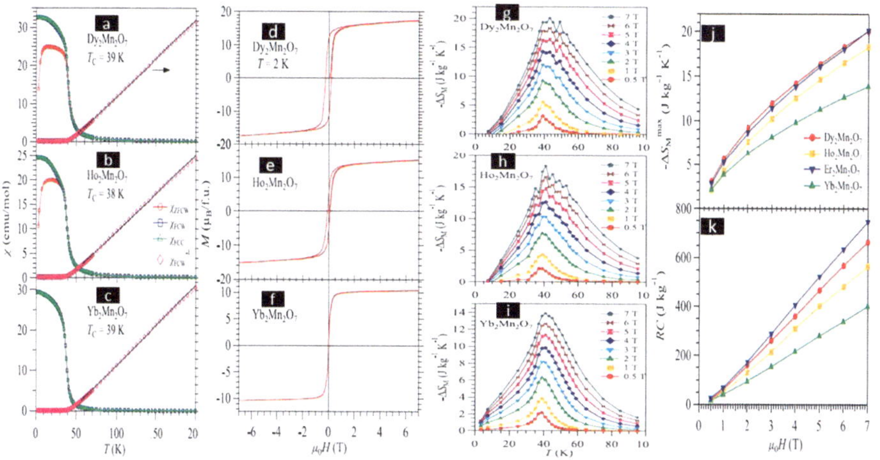

Fig. 6 χ and χ^{-1} curves for **a** $Dy_2Mn_2O_7$, **b** $Ho_2Mn_2O_7$, and **c** $Yb_2Mn_2O_7$ compound performed at H = 0.1 T following the protocol ZFCw, FCw and FCc; χ^{-1} versus T curves is fitted at high temperature using Eq. 2 represented by solid lines. M-H curve recorded at 2 K for **d** $Dy_2Mn_2O_7$, **e** $Ho_2Mn_2O_7$, and **f** $Yb_2Mn_2O_7$. Temperature dependence of ΔS_M curves measured at various field changes up to 7 T for **g** $Dy_2Mn_2O_7$, **h** $Ho_2Mn_2O_7$, and **i** $Yb_2Mn_2O_7$. **j** the maximum of $\Delta {S_M}^{max}$ and **k** RC as a function of field. Adapted from [46]

along with glassy behaviour around 12 K and Griffiths-like state. Figure 8 shows ΔS_M-T curve for $Ga_2Mn_{2-x}Cr_xO_7$ series. The value of $-\Delta S_M$ and RCP are 1.36 J/K-kg and 42.96 J/kg ($x = 0.0$), and 1.47 J/K-kg and 47.73 J/kg ($x = 0.5$). It shows a slight increment in the value of $-\Delta S_M$ and RCP with Cr doping likely due to the mixed valence states of Mn i.e., Mn^{4+} and Mn^{3+}. $-\Delta S_M$ versus T curve shows a minimum at ~10 K emerging due to spin reorientation.

In other recent paper reported by the same group [52], magnetic and MCE properties has been explored in a new pyrochlore compound $Sn_2Mn_2O_7$. It has synthesized by conventional solid-state route at atmospheric pressure. Refinement of XRD pattern suggest monoclinic layered perovskite phase with space group P2/M. The XPS study shows 4+ and 3+ valence state for Sn and Mn, respectively. It undergoes a FM transition at temperature $T_C = 41$ K with SOMPT nature. The estimated value of $-\Delta S_M$ and RCP are found to be 2.94 J/K-kg and 41 J/kg, respectively for the field change of 5 T. It can also be noticed that the value of ΔS_M is slightly higher than what is reported for $Fe_2Mn_2O_7$ and less than $Al_2Mn_2O_7$.

4.2 Ti Based Pyrochlore Oxides

The Pyrochlore oxide $Dy_2Ti_2O_7$ shows spin ice behaviour at low temperature. It obeys ice rule where two spins point outward and tother two spins point inward (2-in/2-out) in each tetrahedron. Experimentally, $Dy_2Ti_2O_7$ shows no phase transition

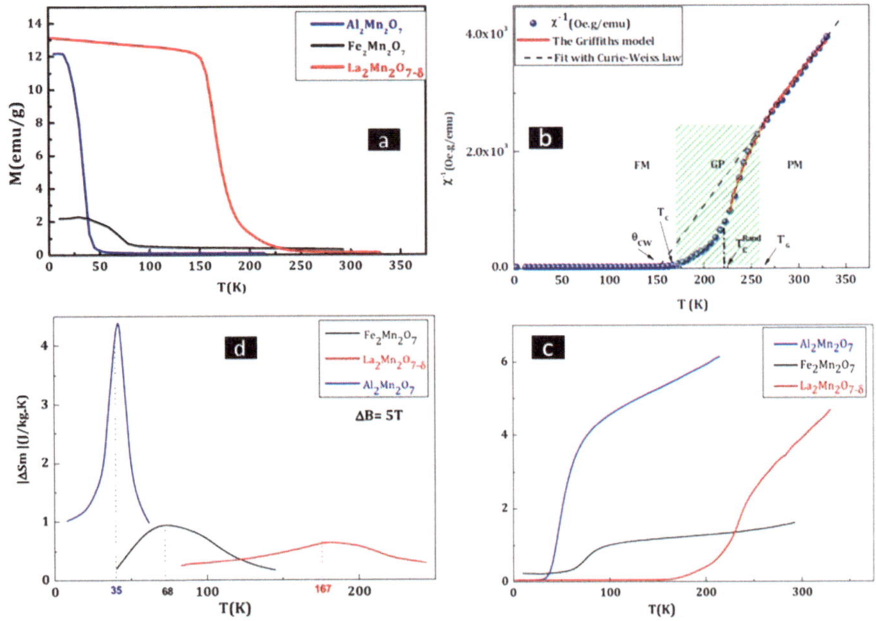

Fig. 7 **a** Temperature dependent dc magnetization, **b** χ^{-1} versus T, **c** χ^{-1} as a function of T and **d** ΔS_M-T curves measured at field changes up to 5 T for the series compounds $A_2Mn_2O_7$ (A = La, Fe, Al). Adapted from [47–50]

at zero magnetic field down to 50 mK. The MCE properties has been explored in $Dy_2Ti_2O_7$ single crystal. Synthesis and characterization can be found elsewhere [53]. Field dependence of ΔS_M has been measured in a magnetic field H $H \parallel$ [111]. The estimated released entropy across the transition found to be 0.5 J/(K-mol-Dy). The peak intensity of S(H) depends strongly on temperature. The peak at $T = 400$, 500 mK is very prominent at the spin-flip field ~ 0.9 T, attributed to the two-in, two-out and the three-in, one-out states. The MCE in Y doped $Dy_2Ti_2O_7$ samples have also been examined at 0.3–6 K in applied field up to 2 T. Polycrystalline powder sample have prepared by well-established solid stat route following the protocol reported elsewhere [54]. It has proposed that in the spin ice regime slow relaxation process may be studied qualitatively using adiabatic demagnetization in the mK range.

Further, a large MCE has been observed in the frustrated pyrochlore AFM $Gd_2Ti_2O_7$ single crystal [55]. In $Gd_2Ti_2O_7$, Gd^{3+} ($4f^7$) exhibit orbital momentum L = 0 and only spin value S = 7/2 ground state. Thus, single ion anisotropy and crystal-field splitting become unimportant. Therefore, system exhibit classical Heisenberg AFM behaviour with dipole–dipole interactions, as a result, system does not show long-range order down to 1 K. The transition below 1 K occurs likely due to dipole–dipole interactions. The values of ΔS_M and RCP at 4.2 K for demagnetization for 90 kOe field change calculated from specific heat data are found to be 17.89 J/K-mol and 30 J/mol-Gd.

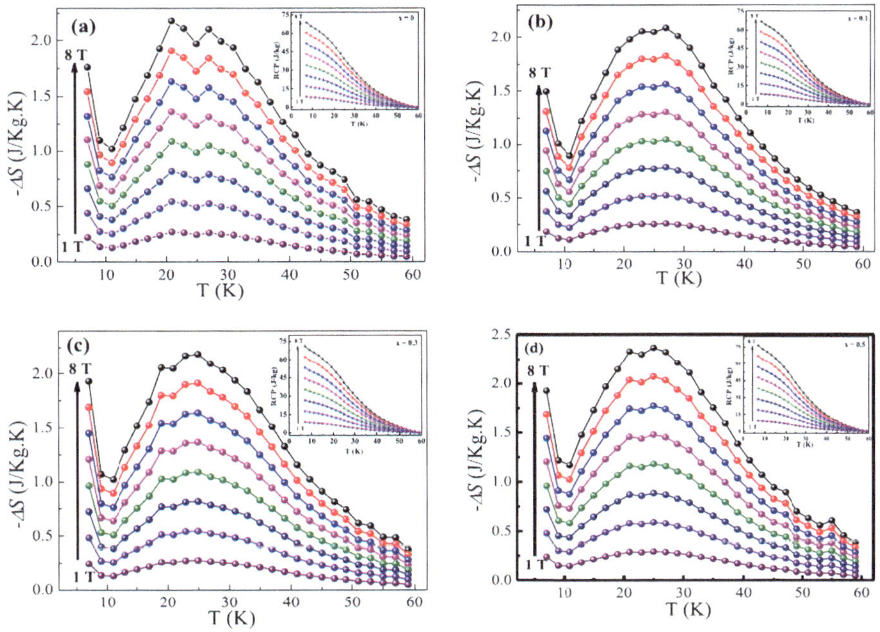

Fig. 8 ΔS_M-T curves measured for the series $Ga_2Mn_{2-x}Cr_xO_7$ (a) $x = 0.0$, **b** $x = 0.1$, **c** $x = 0.3$ and **d** $x = 0.5$; inset demonstrates RCP versus temperature behaviour. Adapted from [51]

The next material of interest in this series is pyrochlore $Er_2Ti_2O_7$. Here, $Er^{3+}(4f^{11})$ ion exhibit single ion anisotropy. Therefore, it can be realized as a frustrated XY AFM system on pyrochlore lattice. It exhibits a large value of negative $\theta_{CW} = -22$ K and displays an AFM transition at ~1.17 K. Figure 10a Displays ΔS_M Versus T Curve of Polycrystalline Er2Ti2O7 Sample Synthesized

By solid state route [56]. It increases with reducing the temperature for a field change of 10 T. It shows symmetrical span at ~ 4.5 K, suggests a second order magnetic phase transition. Figure 10b reveals the value of $\Delta S_M{}^{max}$ and RCP enhances as field increases. Moreover, it has shown that the lowest accessible temperature by demagnetizing single crystal of $Er_2Ti_2O_7$ lies down to 100 mK [57]. This property makes it suitable magnetic coolant above 200 mK.

4.3 TiMn Based Pyrochlore Oxides

The 50% replacement of non-magnetic Ti^{4+} with magnetic Mn^{4+} ion in pyrochlore $Er_2Ti_2O_7$ i.e., Er_2TiMnO_7 compound makes the system interesting. The interaction between Mn^{4+}—$(3d^3)$ and rare earth ion Er^{3+}-$(4f^{11})$ may leads to the complex magnetic properties. Polycrystalline Er_2TiMnO_7 sample synthesized by conventional solid-state route exhibit layered monoclinic structure having space group P2/

the two successive magnetic transition at ~ 10 K and ~ 78 K. It makes the system a potential magnetic coolant in the low temperature range for low field.

4.5 Sn Based Pyrochlore Oxides

$Pr_2Sn_2O_7$ have been reported to exhibit a breakthrough dynamic spin-ice-like state at low temperatures with weak FM interaction $\theta_{CW} = +\ 0.32$ K. Magnetic studies suggested large Ising-like single-ion anisotropy with a non-Kramers doublet and very slow dynamics for the Pr^{3+} (J = 4) crystal field ground states in $Pr_2Sn_2O_7$ pyrochlore. Polycrystalline $Pr_2Sn_2O_7$ sample can be synthesized by conventional solid-state route. The details can be found elsewhere [62]. Figure 13 reveals $-\ \Delta S_M$ versus T curves shows a broad maximum for field change of higher than 4 T between 2 and 3 K suggest the suppression of MCE by the single-ion anisotropy. It is clear that -ΔS_M does not favours an enhancement in MCE up to 15 K despite a strong level of magnetic frustration and change of the applied field is of the order of Tesla. Such behaviour likely due to random orientations of axex with respect to the direction of applied field. Further, the absence of enhanced MCE in a dynamic spin-ice system has supported by theoretical calculation utilizing an approximation of single tetrahedron in a pyrochlore lattice with and without an exchange coupling.

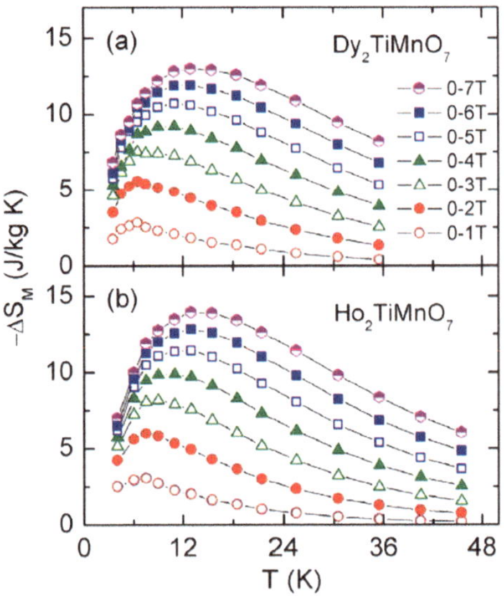

Fig. 11 $-\ \Delta S_M$ versus T curve for **a** Dy_2TiMnO_7, and **b** Ho_2TiMnO_7. Adapted from [59]

5 Ir Based Pyrochlore Oxides $A_2Ir_2O_7$

It have been predicted that the interplay between electronic correlation, spin–orbit coupling, and crystal field effect combined with magnetic frustration in pyrochlore iridates $A_2Ir_2O_7$ (A = rare earth element, Y, Bi) leads to exotic electronic and magnetic phases [63–69]. Therefore, it provides a vast template to explore the complex magnetic phases using the MCE as well. The systematic study of $A_2Ir_2O_7$ shows the enhancement of conductivity with increasing A-site ionic size. For heavier rare earth elements, R = Pr, $Pr_2Ir_2O_7$ shows non-magnetic metallic behaviour. While, an antiferromagnetic insulating-like feature for smaller R = Lu, Yb, Er, Ho, Dy, Tb, and Y was observed. The intermediate candidates R = Eu, Sm, and Nd shows metal insulator transition, associated with the AFM mean field interaction.

Theoretically, $Y_2Ir_2O_7$ has been proposed to show a magnetic Weyl semimetal phase exhibiting an all-in/all out antiferromagnetic (AFM) ground state [63]. Further, MCE have studied in Cr doped $Y_2Ir_2O_7$, i.e., $Y_2Ir_{2-x}Cr_xO_7$ (x = 0.0, 0.05, 0.1 and 0.2) samples [70] and $Er_2Ir_2O_7$ compound [71]. Geometrically frustrated AFM $Y_2Ir_{2-x}Cr_xO_7$ (YICO) series have been reported experimentally to exhibit a Griffiths-phase (GP)-like magnetic behaviour with cluster-glass-like state [69]. Moreover, GP region exhibit Ising-like interaction of spins which provides a hope to achieve high MCE in such frustrated system as it is predicted by theory [26, 27]. Pyrochlore cubic $Y_2Ir_{2-x}Cr_xO_7$ series samples can be synthesized by solid state reaction route.

Figure 14a–c show $-\Delta S_M$-T curves for various change of magnetic fields up to 7 T (for low field see Fig. 14d–f). A positive maximum around T_C can be seen for a moderate value of $-\Delta S_M$ for $x = 0.0$ sample. Surprisingly, $-\Delta S_M$-T becomes negative below T_N for small magnetic field (Fig. 14d). It turns its sign to positive magnetic entropy change and achieve a maximum at T_N as applied magnetic field increases. The shape of $-\Delta S_M$-T curves remain same as it is for the low field. Such MCE features is likely due to a spin-flop-like transition. Inset of Fig. 14d shows a minimum in $-\Delta S_M$-T curve at ~ 5 K, which might be attributed to Ir spin reorientation.

Cr doped samples display the coexistence of conventional MCE and IMCE for high and as well as low field changes (Fig. 14 b, c, e, f). This crossover from high temperature MCE to low temperature IMCE emerges due to the competition between FM and AFM phases likely due to the copresence of Ir^{4+} and Ir^{5+}. It shows significant increment in $-\Delta S_M$ value with substitution of Cr, although the value remains smaller than the standard MCE material Gd.

RC depends on either a large value of $-\Delta S_M$ or a broad δT_{FWHM} of $-\Delta S_M$-T curve or both. Figure 15a shows δT_{FWHM} versus H curve which compensate the small value of MCE, gives rise to a comparable value of RC to the standard MCE materials. The estimated values of RC and RCP around T_N and T_C are displayed in Fig. 15b–d. The important outcomes of YICO series are both the 'high temperatures positive' maximum (near T_C) and 'low temperature negative' minimum (near T_N) in $-\Delta S_M$-T curve extended over a broad range of temperatures. Such coexistence of MCE and IMCE is completely *absent* in the reported pyrochlore oxides. Therefore,

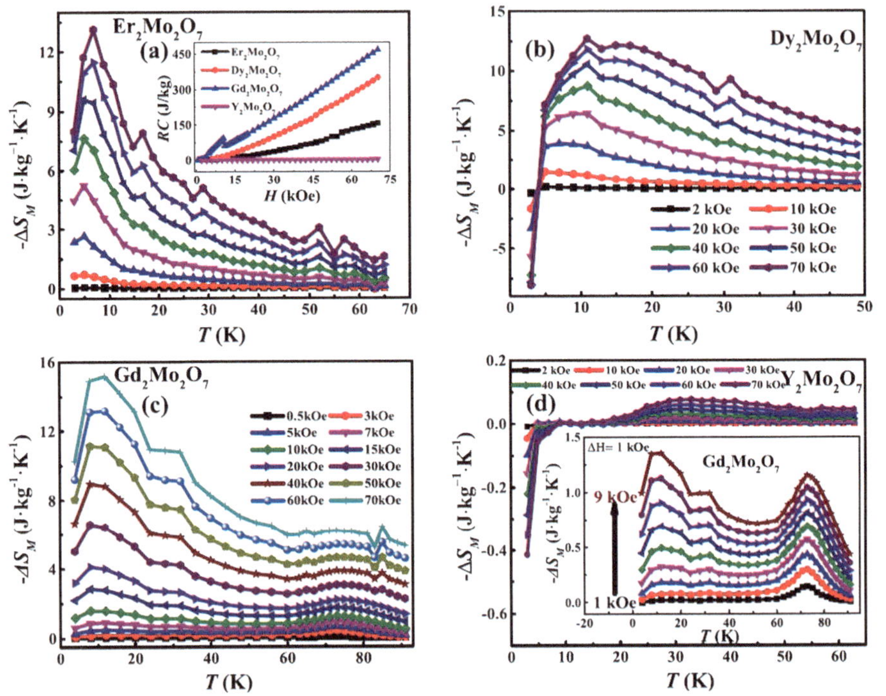

Fig. 12 ΔS_M versus T curve for the field change of 7 T for **a** $Er_2Mo_2O_7$, **b** $Dy_2Mo_2O_7$, **c** $Er_2Mo_2O_7$, and **d** $Y_2Mo_2O_7$. Inset shows RC versus T curves and inset of **d** displays $-\Delta S_M$ versus T curve for $Gd_2Mo_2O_7$ at low field changes. Adapted from [61]

$Y_2Ir_{2-x}Cr_xO_7$ series shows a hope to be in the category of new promising candidates for solid state magnetic cooling.

Moreover, the magnetic properties reported in $Er_2Ir_2O_7$ pyrochlore iridate have shown interesting properties due to the interaction between rare earth element Er^{3+} ($4f^{11}$, S = 3/2, L = 6, J = 15/2) and Ir^{4+} ($5d^5$, S = 1/2, L = 0). $Er_2Ir_2O_7$ shows short range ordering below 2 K for Er-sublattice due to frustration and Er-Ir sublattices drive a weak FM transition at $T_C = 140$ K [72]. Figure 16a shows a broad maximum in $-\Delta S_M$-T curve at $T_C = 140$ K. The maximum obtained of—$\Delta S_M{}^{max}$, and RCP is 3.09 J/K-kg, 128 J/kg, respectively at field change of 5 T (Fig. 16b). It can be noticed that the value of MCE at T_C is larger than what reported for $Y_2Ir_2O_7$ at 5 T field change. It is likely due to the strong magnetization attributed to the interaction between Er^{3+} and Ir^{4+} ion.

Eventually, the important results and parameters are summarized in following Table 1.

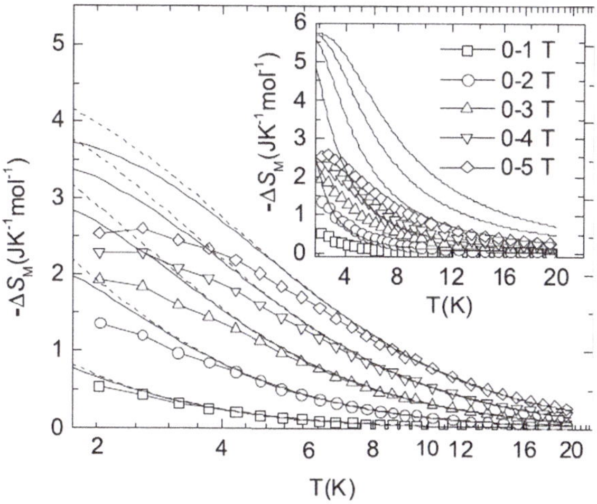

Fig. 13 ΔS_M as a function of T of field changes from 0 to 5 T for compound $Pr_2Sn_2O_7$. The solid and dashed lines represent the theoretical calculations for the involvement of one tetrahedron exhibiting four S = 1/2 Ising spins with and without J/k_B, respectively; J = exchange coupling and k_B = Boltzmann constant. Inset comparison between same data and paramagnetic system with S = 1/2. Adapted from [62]

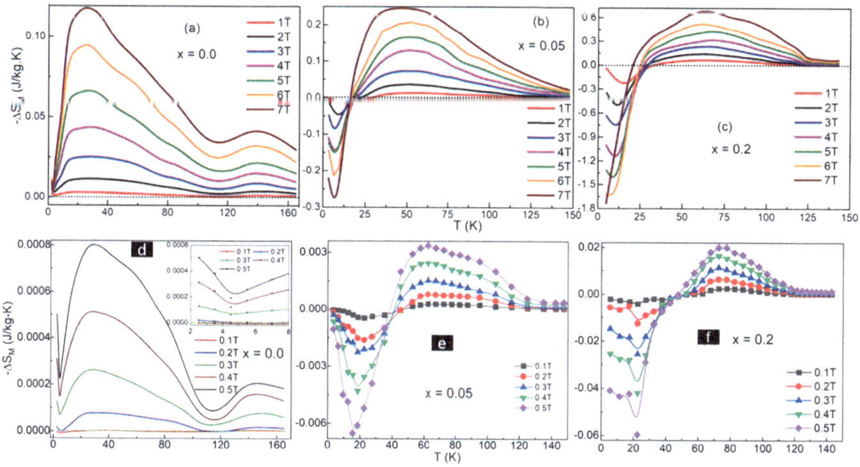

Fig. 14 ΔS_M versus T curves for the $Y_2Ir_{2-x}Cr_xO_7$ series compound for the filed change 1–7 T, **a** $x = 0.0$, **b** $x = 0.05$, **c** $x = 0.2$. Same curve for the field change of 0.1–0.5 T, **d** $x = 0.0$, **e** $x = 0.05$, **f** $x = 0.2$. Adapted from [70]

6 Conclusions

In conclusions, the magnetocaloric effect and magnetic properties of pyrochlore oxides based on Mn, Ti, TiMn, Sn, Mo and Ir materials are reviewed in detail. The crystal structure, magnetic frustration, calculation of MCE parameters is also discussed. The Mn based pyrochlores is reported with various A-site ionic radii with the formula $A_2Mn_2O_7$ (A = La, Dy, Ho, Er, Yb, Sn, Fe, Ga, and Al). Depending on synthesis protocol, $Er_2Mn_2O_7$ crystallizes in two distinct crystal structures i.e.,

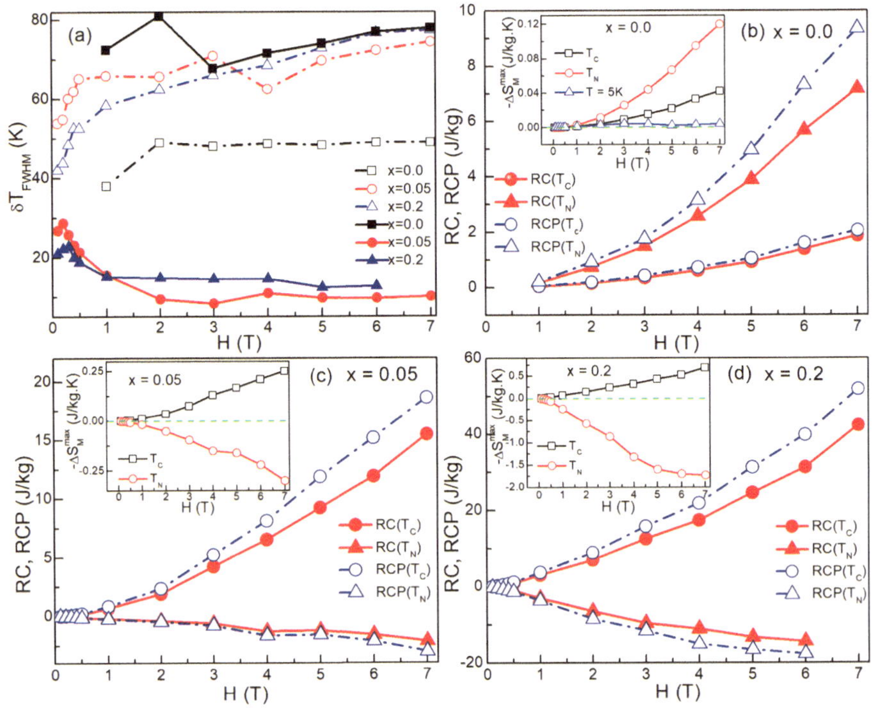

Fig. 15 **a** δ_F WHM versus field H near T_C (hollow symbols) and T_N (solid symbols). Field dependence of RC and RCP for the samples **b** $x = 0.0$, **c** $x = 0.05$, and **d** $x = 0.2$. Inset shows $\Delta S_M{}^{\max}$ versus H curve. Adapted from [70]

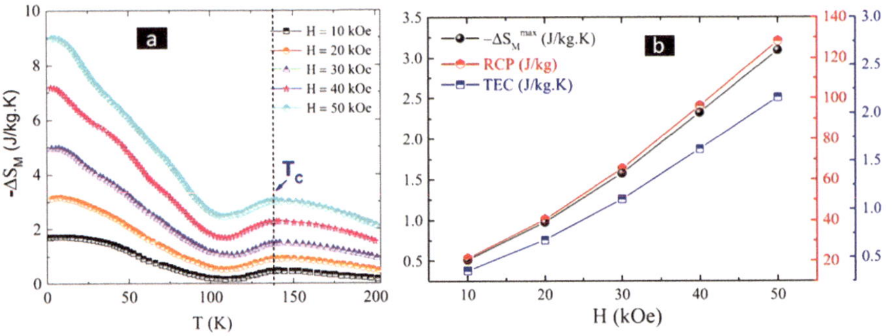

Fig. 16 **a** ΔS_M as a function of T under various field changes upto 5 T for $Er_2Ir_2O_7$ compound. **b** Field dependence of $-\Delta S_M{}^{\max}$, RCP and temperature entropy change (TEC). Adapted from [71]

Table 1 Important MCE parameters estimated for field change of 5 T for various pyrochlore oxides

Sample	Structure	T_C K	T_N K	θ_{CW} K	$(-S_M{}^{max})$ T_C J/K-kg	$(-S_M{}^{max})$ T_N J/K-kg	(RC) T_C J/kg	(RC) T_N J/kg	References
$Er_2Mn_2O_7$	Monoclinic	–	5.47	− 13.92	–	[illegible].37	–	–	[41]
$Er_2Mn_2O_7$	Pyrochlore	35	–	+ 37.4	16.1	–	522	–	[45]
$Dy_2Mn_2O_7$	Pyrochlore	39	–	+ 41.5	15.2	–	515	–	[46]
$Ho_2Mn_2O_7$	Pyrochlore	38	–	+ 39.8	14.0	–	420	–	[46]
$Yb_2Mn_2O_7$	Pyrochlore	38	–	+ 43.5	11.5	–	300	–	[46]
$La_2Mn_2O_7$	Monoclinic	167	–	+ 159	0.65	–	31.5	–	[48, 50]
$Fe_2Mn_2O_7$	Monoclinic	68	–	+ 62	1	–	47.25	–	[47, 50]
$Al_2Mn_2O_7$	Monoclinic	35	–	+ 37	4.5	–	50.25	–	[49, 50]
$Ga_2Mn_2O_7$	Pyrochlore	–	–	− 187.87	–	[illegible].36	–	32.223	[51]
$Ga_2Mn_{1.5}Cr_{0.5}O_7$	Pyrochlore	–	–	− 151.59	–	[illegible].47	–	5.79	[51]
$Sn_2Mn_2O_7$	Monoclinic	41	–	+ 35.97	2.94	–	30.7	–	[52]
$Er_2Ti_2O_7$	Pyrochlore	–	1.17	− 21.54	–	[illegible].77	–	16.6	[56]
Er_2TiMnO_7	Monoclinic	–	2.05	− 20.29	–	[illegible].28	–	–	[58]
Dy_2TiMnO_7	Pyrochlore	7.6	–	+ 10.3	10.73	–	273.9	–	[59]
Ho_2TiMnO_7	Pyrochlore	7.5	–	+ 9.5	11.45	–	267.75	–	[59]
$Er_2Mo_2O_7$	Pyrochlore	–	18	− 12	–	[illegible].6	–	80	[61]
$Dy_2Mo_2O_7$	Pyrochlore	23	–	+ 4.4	10.5	–	230	–	[61]
$Gd_2Mo_2O_7$	Pyrochlore	78	–	+ 66.3	11.1	–	340	–	[61]
$Y_2Mo_2O_7$	Pyrochlore	–	22	− 41.5	–	–	–	–	[61]
$Pr_2Sn_2O_7$	Pyrochlore	–	–	+ 0.32	4.12	–	–	–	[62]
$Er_2Ir_2O_7$	Pyrochlore	140	–	− 26	3.09	–	128	–	[71]

(continued)

Table 1 (continued)

Sample	Structure	T_C K	T_N K	θ_{CW} K	$(-S_M{}^{max})$ T_C J/K-kg	$(-S_M{}^{max})$ T_N J/K-kg	(RC) T_C J/kg	(RC) T_N J/kg	References
$Y_2Ir_2O_7$	Pyrochlore	130	15	− 138	0.022	0.067	0.94	3.9	[70]
$Y_2Ir_{1.8}Cr_{0.2}O_7$	Pyrochlore	70	35	− 140	0.43	1.6	24.7	13.2	[70]

monoclinic layered structure with space group P2/M at ambient pressure and cubic pyrochlore structure with space group $Fd\overline{3}m$ at high pressure. Monoclinic $Er_2Mn_2O_7$ sample undergo an antiferromagnetic (AFM) transition with much lower value of RC (2.37 J/kg), while the $Er_2Mn_2O_7$ sample crystallizes in cubic pyrochlore phase shows ferromagnetic (FM) transition exhibiting higher RC value (522 J/kg) than former for the field change 5 T. Such large reversible MCE in cubic pyrochlore $Er_2Mn_2O_7$ emerges due to two sublattice cooperative FM ordering. Pyrochlore $A_2Mn_2O_7$ series display almost similar MCE behabiour with lower value of RC for $Dy_2Mn_2O_7$, $Ho_2Mn_2O_7$ and $Yb_2Mn_2O_7$ samples as compared to cubic pyrochlore $Er_2Mn_2O_7$. A strong single-ion anisotropy does not allow the full alignment in FM fashion, although Dy, Ho, and Yb exhibit larger magnetic moments than Er. Despite the moderate values of RC for the samples crystallize in monoclinic layered perovskite structure i.e., $La_2Mn_2O_7$, $Fe_2Mn_2O_7$ and $Al_2Mn_2O_7$, the MCE values and strength of Griffiths-like state enhances with reduction of A-site ionic radius. In cubic pyrochlore $Ga_2Mn_2O_7$ sample doped with magnetic Cr at magnetic Mn-site, i.e., $Ga_2Mn_{2-x}Cr_xO_7$ (x = 0, 0.1, 0.3, 0.5), the value of RC shows enhancement with Cr doping likely due to mixed oxidation states of Mn^{3+} and Mn^{4+}. The series also shows the presence of Griffiths-like state along with glassy phase. The $Sn_2Mn_2O_7$ compound crystallizes in monoclinic layered perovskite structure undergoes FM transition exhibiting moderate RC value. XPS study has suggested 4 + and 3 + valence state of Sn and Mn, respectively, which is *surprising*.

The titanium-based sample $Er_2Ti_2O_7$ crystallized in cubic pyrochlore structure shows AFM transition with low value of RC likely due to single-ion anisotropy which prevents long range ordering. The replacement of 50% nonmagnetic Ti atom with magnetic Mn atom i.e., Er_2TiMnO_7 acquires monoclinic structure exhibit moderate value of reversible MCE just above the boiling temperature of helium. The pyrochlore cubic compounds Dy_2TiMnO_7 and Ho_2TiMnO_7 shows FM transition exhibit significant enhancement in RC value.

The molybdate series $Er_2Mo_2O_7$, $Dy_2Mo_2O_7$, $Gd_2Mo_2O_7$ and $Y_2Mo_2O_7$ crystallizes in pyrochlore structure show spin-glass transition for the samples A = Er, Dy, Y while it undergoes FM transition and subsequently exhibits a reentrant spin-glass behavior for A = Gd sample. The ob- tained RC value is significantly larger in $Gd_2Mo_2O_7$ compound while $Y_2Mo_2O_7$ exhibit negligible MCE value as compared to the other members of the series $A_2Mo_2O_7$.

Iridium based compounds exhibit the interplay of spin-orbit coupling, electronic correlations, crystal electric field and magnetic frustration which provide a broad window to explore the complex magnetic and electronic states. $Er_2Ir_2O_7$ undergoes a weak ferromagnetic transition likely due to the ordering of Ir^{4+} moments. It exhibits significant value of RC likely arises due to the interaction of rare earth element Er and transition metal Ir.

Eventually, the pyrochlore YICO series undergo various magnetic transitions, which is likely attributed to the coexistence of FM and AFM clusters induced by frustration. It gives rise to the IMCE at low temperature and conventional MCE at high temperature. In this series, substitution of Ir with Cr leads to not only coexistence of inverse MCE and conventional MCE but order of magnitude increment of $-\Delta S_M$.

Although the value of RC is not large, the conventional MCE and IMCE spread over a giant working temperature window. It can be noticed that coexistence of IMCE and MCE is completely *absent* in all reported pyrochlore oxides except YICO series.

Besides a detail review of all reported litterateurs of MCE in pyrochlore oxides, it is interesting to see that $A_2Mn_2O_7$ series crystallizes in monoclinic layered perovskite structure and cubic pyrochlore phase depending on synthesis process. Therefore, more experiments on various com- pounds are required to shed the light on this point. Hopefully, this article may open up a future direction to encourage the researchers working in various fields associated to the MCE materials.

References

1. Ayas, A.O., Etin, S.K., Aka, G., Akyol, M., Ekicibil, A.: Magnetic refrigeration: current progress in magnetocaloric properties of perovskite manganite materials. Mater. Today Commun. **35**, 105988 (2023)
2. Brown, J.S., Domanski, P.A.: Review of alternative cooling technologies. Appl. Therm. Eng. **64**, 252 (2014)
3. Tishin, A.M., Spichkin, Y.I.: The magnetocaloric effect and its applications, CRC Press; 1st edition, (2003)
4. Tishin, A.M., Spichkin, Y.I., Zverev, V.I., Egolf, P.W.: A review and new perspectives for the magnetocaloric effect: new materials and local heating and cooling inside the human body. Int. J. Refrig. **68**, 177 (2016)
5. Phan, M.-H., Yu, S.-C.: Review of the magnetocaloric effect in manganite materials. J. Magn. Magn. Mater. **308**, 325 (2007)
6. Kitanovski, A.: Energy applications of magnetocaloric materials. Adv. Energy Mater. **10**, 1903741 (2020)
7. Lyubina, J.: Magnetocaloric materials for energy efficient cooling. J. Phys. D Appl. Phys. **50**, 053002 (2017)
8. Franco, V., Blazquez, J.S., Ipus, J.J., Law, J.Y., Moreno-Ramirez, L.M., Conde, A.: Magnetocaloric effect: from materials research to refrigeration devices. Prog. Mater. Sci. **93**, 112 (2018)
9. Liu, J., Gottschall, T., Skokov, K.P., Moore, J.D., Gutfleisch, O.: Giant magnetocaloric effect driven by structural transitions. Nat. Mater. **11**, 620 (2012)
10. Liu, P., Yuan, D., Dong, C., Lin, G., Villora, E.G., Qi, J., Zhao, X., Shimamura, K., Ma, J., Wang, J., Zhang, Z., Li, B.: Ultralow-field magnetocaloric materials for compact magnetic refrigeration. NPG Asia Mater. **15**, 41 (2023)
11. Shen, B.G., Sun, J.R., Hu, F.X., Zhang, H.W., Cheng, Z.H.: Recent progress in exploring magnetocaloric materials. Adv. Mater. **21**, 4545 (2009)
12. Pecharsky, V.K., Gschneidner, K.A., Jr.: Giant magnetocaloric effect in $Gd_5(Si_2Ge_2)$. Phys. Rev. Lett. **78**, 4494 (1997)
13. Moya, X., Hueso, L.E., Maccherozzi, F., Tovstolytkin, A.I., Podyalovskii, D.I., Ducati, C., Phillips, L.C., Ghidini, M., Hovorka, O., Berger, A., Vickers, M.E., Defay, E., Dhesi, S.S., Mathur, N.D.: Giant and reversible extrinsic magnetocaloric effects in $La_{0.7}Ca_{0.3}MnO_3$ films due to strain. Nat. Mater. **12**, 52 (2013)
14. Guillou, F., Pathak, A.K., Paudyal, D., Mudryk, Y., Wilhelm, F., Rogalev, A., Pecharsky, V.K.: Non- hysteretic first-order phase transition with large latent heat and giant low-field magnetocaloric effect. Nat. Commun. **9**, 2925 (2018)
15. Dankov, S.Y., Tishin, A.M., Pecharsky, V.K., Gschneidner, K.A., Jr.: Magnetic phase transitions and the magnetothermal properties of gadolinium. Phys. Rev. B **57**, 3478 (1998)

16. Krenke, T., Duman, E., Acet, M., Wassermann, E.F., Moya, X., Manosa, L., Planes, A.: Inverse magnetocaloric effect in ferromagnetic NiMnSn alloys. Nat. Mater. **4**, 450 (2005)
17. Zhang, X., Zhang, B., Yu, S., Liu, Z., Xu, W., Liu, G., Chen, J., Cao, Z., Wu, G.: Combined giant inverse and normal magnetocaloric effect for room-temperature magnetic cooling. Phys. Rev. B **76**, 132403 (2007)
18. Biswas, A., Chandra, S., Samanta, T., Ghosh, B., Datta, S., Phan, M.H., Raychaudhuri, A.K., Das, I., Srikanth, H.: Universality in the entropy change for the inverse magnetocaloric effect. Phys. Rev. B **87**, 134420 (2013)
19. Alman, V., Annadi, A., Khan, M.A., Singh, V., Bohra, M.: Coexistence of normal and inverse magnetocaloric effect in inhomogeneous $Ni_{95}Cr_5$ layers. Phys. Scr. **98**, 095907 (2023)
20. Bohra, M., Singh, N., Annadi, A., Battula, S.V., Singh, V.: Leveraging of both positive and negative magnetocaloric effects in $ZnFe_2O_4$ layers. Mater. Res. Bull. **169**, 112547 (2024)
21. Tegus, O., Bruck, E., Buschow, K.H.J., de Boer, F.R.: Transition-metal-based magnetic refrigerants for room-temperature applications. Nature **415**, 150 (2002)
22. Chaudhary, V., Chen, X., Ramanujan, R.V.: Iron and manganese based magnetocaloric materials for near room temperature thermal management. Prog. Mater. Sci. **100**, 64 (2019)
23. Biswas, A., Pathak, A.K., Zarkevich, N.A., Xubo, L., Mudryk, Y., Balema, V., Johnson, D.D., Pecharsky, V.K.: Designed materials with the giant magnetocaloric effect near room temperature. Acta Mater. **180**, 341 (2019)
24. Waske, A., Gruner, M.E., Gottschall, T., Gutfleisch, O.: Magnetocaloric materials for refrigeration near room temperature. MRS Bull. **43**, 269 (2018)
25. Alman, V., Ramakrishna, V.V., Battula, S.V., Sahu, B.N., Annadi, A., Singh, V., Bohra, M.: Exploration of room-temperature magnetocaloric effect in nanogranular Ni-Cr thin films. MRS Commun. **13**, 117 (2023)
26. Zhitomirsky, M.E.: Enhanced magnetocaloric effect in frustrated magnets. Phys. Rev. B **67**, 104421 (2003)
27. Jurcisinova, E., Jurcisin, M.: Adiabatic cooling processes in frustrated magnetic systems with pyrochlore structure. Phys. Rev. E **96**, 052128 (2017)
28. Kennedy, B.J.: Structural trends in pyrochlore-type oxides. Physica B Condens. Matter. **241**, 303 (1997)
29. Gardner, J.S., Gingras, M.J.P., Greedan, J.E.: Magnetic pyrochlore oxides. Rev. Mod. Phys. **82**, 53 (2010)
30. Ramirez, A.P.: Strongly geometrically frustrated magnets. Annu. Rev. Mater. Sci. **24**, 453 (1994)
31. Binder, K., Young, A.P.: Spin glasses: Experimental facts, theoretical concepts, and open questions. Rev. Mod. Phys. **58**, 801 (1986)
32. Griffiths, R.B.: Nonanalytic behavior above the critical point in a random ising ferromagnet. Phys. Rev. Lett. **23**, 17 (1969)
33. Gingras, M. J. P., Stager, C. V., Raju, N. P., Gaulin, B. D., and Greedan, J. E.: Static critical behavior of the spin-freezing transition in the geometrically frustrated pyrochlore antiferromagnet $Y_2Mo_2O_7$. Phys. Rev. Lett. 78, 947 (1997)
34. Takatsu, H., Kadowaki, H., Sato, T.J., Lynn, J.W., Tabata, Y., Yamazaki, T., Matsuhira, K.: Quantum spin fluctuations in the spin-liquid state of $Tb_2Ti_2O_7$. J. Phys. Condens. Matter **24**, 052201 (2012)
35. Bramwell, S.T., Gingras, M.J.P.: Spin ice state in frustrated magnetic pyrochlore materials. Science **294**, 1495 (2001)
36. Petit, S., Bonville, P., Mirebeau, I., Mutka, H., Robert, J.: Spin dynamics in the ordered spin ice $Tb_2Sn_2O_7$. Phys. Rev. B **85**, 054428 (2012)
37. Taguchi, Y., Oohara, Y., Yoshizawa, H., Nagaosa, N., Tokura, Y.: Spin chirality, berry phase, and anoma-lous hall effect in a frustrated ferromagnet. Science **291**, 2573 (2001)
38. Yasui, Y., Kageyama, T., Moyoshi, T., Soda, M., Sato, M., Kakurai, K.: Studies of anomalous hall effect and magnetic structure of $Nd_2Mo_2O_7$-test of chirality mechanism. J. Phys. Soc. Jpn. **75**, 084711 (2006)

39. Hanawa, M., Muraoka, Y., Tayama, T., Sakakibara, T., Yamaura, J., Hiroi, Z.: Superconductivity at 1 K in $Cd_2Re_2O_7$. Phys. Rev. Lett. **87**, 187001 (2001)
40. Nakatsuji, S., Machida, Y., Maeno, Y., Tayama, T., Sakakibara, T., van Duijn, J., Balicas, L., Millican, J.N., Macaluso, R.T., Chan, J.Y.: Metallic spin-liquid behavior of the geometrically frustrated kondo lattice $Pr_2Ir_2O_7$. Phys. Rev. Lett. **96**, 087204 (2006)
41. Ghosh, B., Dwivedi, V.K., Mukhopadhyay, S.: Non-Fermi liquid regime in metallic pyrochlore iridates: quantum Griffiths singularities. Phys. Rev. B **102**, 144444 (2020)
42. Amor, N.B., Bejar, M., Dhahri, E., Valente, M.A., Garden, J.L., Hlil, E.K.: Magnetocaloric effect in the vicinity of second order antiferromagnetic transition of $Er_2Mn_2O_7$ compound at different applied magnetic field. J. Alloys Compd. **563**, 28 (2013)
43. Subramanian, M.A., Aravamudan, G., Subba Rao, G.V.: Oxide pyrochlores–a review. Prog. Solid State Chem. **15**, 55 (1983)
44. Kim, W.S., Yang, J.-K., Lee, C.-K., Lee, H.-S., Park, H.-H.: Synthesis and characterization of ferroelectric properties of $Ce_2Ti_2O_7$ thin films with Ce^{3+} by chemical solution deposition. Thin Solid Films **517**, 506 (2008)
45. Cai, Y.Q., Jiao, Y.Y., Cui, Q., Cai, J.W., Li, Y., Wang, B.S., Fernandez-Diaz, M.T., McGuire, M.A., Yan, J.-Q., Alonso, J.A., Cheng, J.-G.: Giant reversible magnetocaloric effect in the pyrochlore $Er_2Mn_2O_7$ due to a cooperative two-sublattice ferromagnetic order. Phys. Rev. Mater. **1**, 064408 (2017)
46. Cui, Q., Wang, N.N., Su, N., Cai, Y.Q., Wang, B.S., Shinmei, T., Irifune, T., Alonso, J.A., Cheng, J.G.: Large reversible magnetocaloric effect in the ferromagnetic pyrochlores $R_2Mn_2O_7$ (R = Dy, Ho, Yb). J. Magn. Magn. Mater. **490**, 165494 (2019)
47. Khachnaoui, F., Amor, N.B., Bejar, M., Dhahri, E., Hlil, E.K.: Synthesis and Magnetic Properties of New Pyrochlore $Fe_2Mn_2O_7$ Compound. J. Supercond. Nov. Magn. **31**, 3803 (2018)
48. Khachnaoui, F., Amor, N.B., Bejar, M., Dhahri, E., Hlil, E.K.: Appearance of Griffiths-like phase in a new pyrochlore compound $La_2Mn_2O_{7-\delta}$. J. Supercond. Nov. Magn. **32**, 2133 (2019)
49. Khachnaoui, F., Amor, N.B., Nouri, K., Bejar, M., Dhahri, E.: Investigation of Griffiths-like phase at low temperature in a new magnetocaloric compound, $Al_2Mn_2O_7$. J. Phys. Chem. Solids **148**, 109605 (2021)
50. Khachnaoui, F., Amor, N.B., Nouri, K., Bejar, M., Dhahri, E.: Ionic radius effect on the evolution of the Griffiths phase and the magnetocaloric response at low temperature of $A_2Mn_2O_7$ (A: Fe, Al and La). Solid State Commun. **360**, 115031 (2023)
51. Singh, K., Panwar, N.: Observation of Griffiths phase, spin glass behaviour and magnetocaloric effect in frustrated $Ga_2Mn_{2-x}Cr_xO_7$ (x = 0, 0.1, 0.3, 0.5) pyrochlore compounds. Curr. Appl. Phys. **53**, 86 (2023)
52. Singh, K., Rani, M., Malik, V.K., Panwar, N.: Structural, optical, magnetic, and magnetocaloric properties of $Sn_2Mn_2O_7$ pyrochlore. Phys. Status Solidi A **221**, 2300553 (2024)
53. Aoki, H., Sakakibara, T., Matsuhira, K., Hiroi, Z.: Magnetocaloric effect study on the pyrochlore spin ice compound $Dy_2Ti_2O_7$ in a [111] magnetic field. J. Phys. Soc. Jpn. **73**, 2851 (2004)
54. Orendac, M., Hanko, J., Cizmar, E., Orendacova, A., Shirai, M., Bramwell, S.T.: Magnetocaloric study of spin relaxation in dipolar spin ice $Dy_2Ti_2O_7$. Phys. Rev. B **75**, 104425 (2007)
55. Sosin, S.S., Prozorova, L.A., Smirnov, A.I., Golov, A.I., Berkutov, I.B., Petrenko, O.A., Balakrishnan, G., Zhitomirsky, M.E.: Magnetocaloric effect in pyrochlore antiferromagnet $Gd_2Ti_2O_7$. Phys. Rev. B **71**, 094413 (2005)
56. Amor, N.B., Bejar, M., Hussein, M., Dhahri, E., Valente, M.A., Hlil, E.K.: Synthesis, magnetic properties, magnetic entropy and arrot plot of antiferromagnetic frustrated $Er_2Ti_2O_7$ compound. J. Supercond. Nov. Magn. **25**, 1035 (2012)
57. Wolf, B., Tutsch, U., Dorschug, S., Krellner, C., Ritter, F., Assmus, W., Lang, M.: Magnetic cooling close to a quantum phase transition—the case of $Er_2Ti_2O_7$, J. Appl. Phys. **120**, 142112 (2016)
58. Amor, N. B., Bejar, M., Dhahri, E., Valente, M. A., Garden, J. L., and Hlil, E. K.: Magnetic and magnetocaloric properties of Er_2TiMnO_7 compound. J. Supercond. Nov. Magn. **26**, 3455 (2013)

59. Zhang, Y., Li, H., Guo, D., Ren, Z., and Wilde, G.: Cryogenic magnetic properties in the pyrochlore RE_2TiMnO_7 (RE = Dy and Ho) compounds. Ceram. Int. **44**, 15681 (2018)
60. Hanasaki, N., Watanabe, K., Ohtsuka, T., Kzsmarki, I., Iguchi, S., Miyasaka, S., and Tokura, Y.: Nature of the transition between a ferromagnetic metal and a spin-glass insulator in pyrochlore molybdates. Phys. Rev. Lett. **99**, 086401 (2007)
61. Wu, Y.-D., Dong, Q.-Y., Ma, Y., Ke, Y.-J., Su, N., Zhang, X.-Q., Wang, L.-C., and Cheng, Z. H.: Phase transition-induced magnetocaloric effects in $R_2Mo_2O_7$ (R = Er, Dy, Gd and Y) molybdates. Mater. Lett. **198**, 110 (2017)
62. Tkac, V., Tarasenko, R., Tothova, E., Bujnakova, Z., Tibenska, K., Orendacova, A., Sechovsky, V., Orendac, M.: Relaxation phenomena and magnetocaloric effect in the dynamic spin ice $Pr_2Sn_2O_7$. J. Alloys Compd. **808**, 151719 (2019)
63. Krempa, W.W., Chen, G., Kim, Y.B., Balents, L.: Correlated quantum phenomena in the strong spin-orbit regime. Annu. Rev. Condens. Matter Phys. **5**, 57 (2014)
64. Juyal, A., Dwivedi, V.K., Verma, S., Nandi, S., Agarwal, A., Mukhopadhyay, S.: Possible transition between charge density wave and Weyl semimetal phase in $Y_2Ir_2O_7$. Phys. Rev. B **106**, 155149 (2022)
65. Dwivedi, V.K., Mukhopadhyay, S.: Suppression of long range magnetic ordering and electrical conduction in $Y_{1.7}Bi_{0.3}Ir_2O_7$ thin film, J. Magn. Magn. Mater. **484**, 313 (2019). Evolution of structural, magnetic and electrical transport properties of $Y_{1.7}Bi_{0.3}Ir_2O_7$ thin film grown on YSZ(100) substrate by PLD. Physica B: Condens. Matter **571**, 137 (2019)
66. Dwivedi, V.K., Mukhopadhyay, S.: Coexistence of high electrical conductivity and weak ferromagnetism in Cr doped $Y_2Ir_2O_7$ pyrochlore iridates. J. Appl. Phys. **125**, 223901 (2019). Influence of electronic structure parameters on the electrical transport and magnetic properties of $Y_{2-x}Bi_xIr_2O_7$ pyrochlore iridates. J. Appl. Phys. **126**, 165112 (2019)
67. Dwivedi, V.K., Juyal, A., Mukhopadhyay, S.: Colossal enhancement of electrical conductivity in $Y_2Ir_2O_7$ nanoparticles. Mater. Res Express **3**, 115020 (2016). Structural and magnetic studies of nanocrystalline $Y_2Ir_2O_7$, AIP Conf. Proc. **1665**, 050160 (2015)
68. Dwivedi, V.K., Mukhopadhyay, S.: Effect of stoichiometry on magnetic and transport properties in polycrystalline $Y_2Ir_2O_7$. Conf. A.I.P.: Proc. **1953**, 120002 (2018). Study of optical properties of polycrystalline $Y_2Ir_2O_7$. AIP Conf. Proc. **1832**, 090016 (2017)
69. Dwivedi, V.K.: Identification of a Griffiths-like phase and its evolution in the substituted pyrochlore iridates $Y_2Ir_{2-x}Cr_xO_7$ (x = 0.0, 0.05, 0.1, 0.2). Phys. Rev. B **107**, 134413 (2023)
70. Dwivedi, V.K., Mandal, P., Mukhopadhyay, S.: Frustration induced inversion of magnetocaloric effect and metamagnetic transition in substituted pyrochlore iridates. Appl. A.C.S.: Electron. Mater. **4**, 1611 (2022)
71. Ezairi, S., Elouafi, A., Lmai, F., Tizliouine, A.: Structural, magnetic and magnetocaloric effect of pyrochlore iridate $Er_2Ir_2O_7$. Phys. Scr. **98**, 065803 (2023)
72. Lefranois, E., Simonet, V., Ballou, R., Lhotel, E., Hadj-Azzem, A., Kodjikian, S., Lejay, P., Manuel, P., Khalyavin, D., Chapon, L.C.: Anisotropy-tuned magnetic order in pyrochlore iridates. Phys. Rev. Lett. **114**, 247202 (2015)

Plasmonic Nanoparticles and Nanostructures for High Performance Optoelectronic Applications

Avesh Kumar, Anju Mishra, Vishakha Kaushik, and Amodini Mishra

Abstract Noble metal nanoparticles corresponding to surface plasmon resonance wavelength are excellent light absorbers. The absorption processes create highly energetic electron (hot electrons) near or on the surface of plasmonic nanoparticles. These energetic electrons enhance the local electric field on the metal nanoparticles that leads to the formation of an electromagnetic wave. This wave strongly polarizes the atoms of the molecule and increase electrons vibration of the atoms. As the electrons vibration increase, the induced dipole moment oscillates at the wave frequency. The induced dipole moment will affect velocity, density of electrons of each nanoparticle which will improve optical property propagating light through a nonlinear medium. The optical property of the material can be tuned by controlling shape and size of nanostructure. The tunable properties play a major role in optical applications such as biosensors, photocatalysis, photovotatic devices, laser technology, energy store devices and other optical applications. Thus the noble metal nanoparticles, whose wavelength corresponds to the wavelength of interact light, have vast application in science and technology.

Keywords Noble metal nanoparticles · Surface plasmon resonance · Laser beam · Electric field · Optical nonlinearity · Far-field diffraction rings

---

Avesh Kumar, Anju Mishra, Vishakha Kaushik, These authors have equal contribution to work

---

A. Kumar
Physical Research Laboratory, Navrangpura, Ahmedabad, India

A. Mishra
Department of Chemistry, Indian Institute of Technology Kanpur, Kanpur, U.P, India

V. Kaushik
Materials and Nano Engineering Research Laboratory, Department of Physics, School of Physical Sciences, DIT University, Dehradun, India

A. Mishra (✉)
Department of Physics, Indian Institute of Technology Delhi, Hauz Khas, New Delhi, India
e-mail: amodiniphysics@gmail.com

A. Mishra and V. Pathak (eds.), *Smart Materials for Energy Storage and Biomedical Applications*, Engineering Materials, https://doi.org/10.1007/978-3-031-92584-9_3

1 Introduction

The nonlinear optics is one of the important disciplines optical physics in which light interact with nonlinear materials. The nonlinear optics has new phenomena and new effects in nonlinear medium. The optical properties of the material can be modified by interaction of laser light with the nonlinear optical material [1–3]. Typically, only laser light is sufficiently intense to generate the nonlinear optical phenomena in the material. Nonlinear optical properties promote the design and fabrication of optoelectronic and photonic devices. Nonlinear optics is widely used in optical computing, storage, laser technology, information, light aspects of communication and image processing. The enhancement of the nonlinear optical response with metal nanoparticles is often associated surface plasmon excitations that are collective electromagnetic modes. Therefore plasmonic metal nanoparticles and dielectric materials can be effectively used in optoelectronic applications [4–7].

Noble metal nanoparticles have wide attraction because of their unique properties such as electronic, plasmonic, photonics, catalysis, biomedical and others [8–10]. The interaction of laser light with plasmonic nanoparticles whose wavelength corresponding to the wavelength of light, have new physics and photovoltaic, photocatalysis, laser technology, photonic and other optoelectronics applications. The plasmonic nanoparticles are commonly used for improving the sensing performance of surface plasmon resonance sensors. The plasmonic nanoparticles exhibit strong absorption corresponding to excitation wavelength of light. Consequently, surface plasmon resonance produce a large local electric field on the surface of the nanoparticles and enhance nonlinear absorption. It has been exhibited that the surface plasmon resonance sensitivity is closely related to the excited electric field [11, 12]. Therefore larger the electric field has more sensitive of surface plasmon resonance sensors to the change of its surrounding medium. When the light of wavelength corresponding to surface plasmon resonance absorption propagates through metal nanoparticles strongly absorb visible light due to their surface plasmon resonance.

The noble metal nanoparticles that are supported on a metal oxide surface were found to be highly efficient for nonlinear optical applications. Among all noble metal nanoparticles, Au nanoparticles supported on TiO_2 have found maximum applications such as photocatalysis, photovoltaic cell, biomedical science, optoelectronics, energy storage devices and etc. The electrons transfer from gold nanoparticles surface to surrounding TiO_2 semiconductor significantly modifies optical properties of Au-TiO_2 nanocomposites [13–15]. The photo excited electrons transfer from Au nanoparticles to TiO_2 semiconductor. The transfer of electron can change in work function of Au-TiO_2 nanocomposites. That is, Fermi level of Au-TiO_2 nanocomposites shifts with concentration of Au. The Fermi energy level of nanocomposites depends on density of accumulated electrons in the conduction band of semiconductor, these accumulated electrons is associated to the enhancement of local electric field near interface of Au-TiO_2 nanocomposites by surface Plasmon resonance. This field enhancement is responsible for the large enhancement in the optical nonlinearity.

This incensement in optical nonlinearity can play a key role in the development of materials for optoelectronic and photonic application [16–18].

2 Noble Metal Nanoparticles

The absorption spectrum of gold nanoparticles is given in Fig. 1. The absorption spectrum consists of a band at wavelength 522 nm corresponding to the surface plasmon resonance of gold nanoparticles. This shows formation of Au nanoparticles. The absorption spectrum associates with the enhancement of the electromagnetic field within metal nanoparticles. The Au atom forms the nuclei for metal nanoparticles by surface plasma of the free electron of the conduction band. These Au nuclei further grow to form gold nanocluster. The surface plasmon resonance absorption occurs due to the oscillation of free electron near the surface of plasmon nanoparticles. The surface plasmon resonance of nanoparticles strongly depends on the conduction band electron concentrations [19, 20]. For metal nanoparticles, surface plasmon resonance occurs at a frequency of $\omega_{SPR} = \omega_p/\sqrt{3}$, where the plasma frequency ω_p directly depends on the electrons concentration n_e by the relation.

$$\omega_p = \sqrt{\frac{n_e e^2}{m^* \varepsilon_0}}$$

where e is the charge of an electron, m^* is the effective mass of an electron and ε_0 is the permittivity of free space.

The interaction of electromagnetic radiation with noble metal nanoparticles excites to the electrons of the metal nanoparticles which oscillate with certain frequency. When the frequency of oscillate electrons matches with the frequency

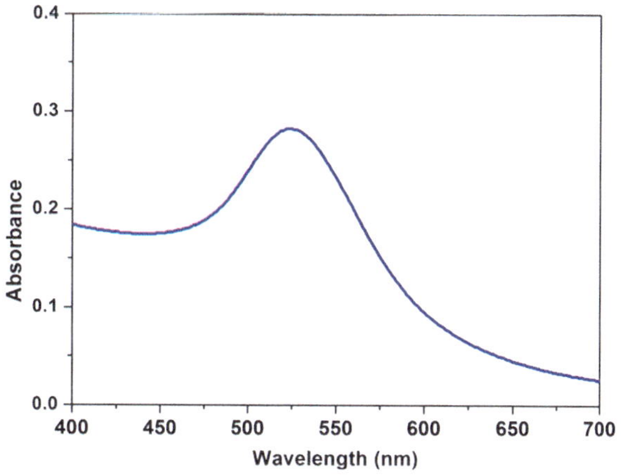

Fig. 1 UV–visible spectrum of Au nanoparticles

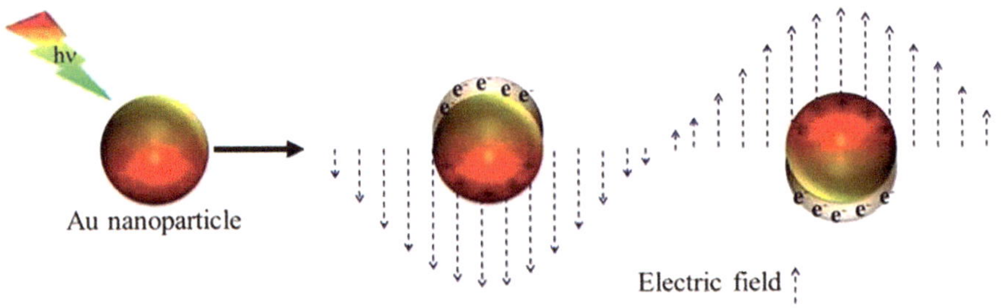

Fig. 2 The conduction electrons oscillating across the Au nanoparticle in the electric field of incident light

of electromagnetic wave, the light is absorbed by the noble metal nanoparticles. At which frequency light is absorbed, is called resonance frequency and this phenomenon is called surface plasmon resonance [21–23]. The surface plasmon resonance occurs near UV–visible region for the noble metals (such as Au, Ag, etc.) nanoparticles. The plasmon resonance wave travels along the interface between noble metal and dielectric that is called surface plasmon polaritons. On other hand, collective excitation of conduction band electron in noble metal particle is called localized surface plasmon resonance (as shown in Fig. 2). Its frequency can be tuned by the changing shape, size, surrounding medium and the material of the noble metal nanoparticles. The localized surface plasmon resonance enhances local electric field near and on the surface of metal nanoparticles. The huge electric field on the particle surface strongly polarizes the atoms of the molecule [24, 25].

The polarized atoms induce polarization nearby or at the interface between metallic nanoparticle and dielectric material that can change the restoring force of the electronic cloud. Thereby changes the frequency of the electrons oscillation that causes red or blue shift in the absorption peak [26].

When the particle size of the materials becomes comparable to the wavelength of the electromagnetic radiation than the electrostatics is not valid. In this case the problem can be solved using electrodynamics. When the particle size is greater than the wavelength of electromagnetic radiation then the incident electric field over the particle does not remain same at any time. By which, the collective excitation of electron clouds does not remain in phase within particle. Consequently, in noble metal nanoparticles [27, 28], a red shift in plasmonic resonance and broadening of the absorption peak occurs. This phenomenon is called dynamic depolarization.

2.1 *Interaction of Laser Light with Au Nanoparticles*

When the laser light corresponding to the surface plasmon resonance absorption propagates through a nonlinear medium, metal nanoparticles strongly absorb laser

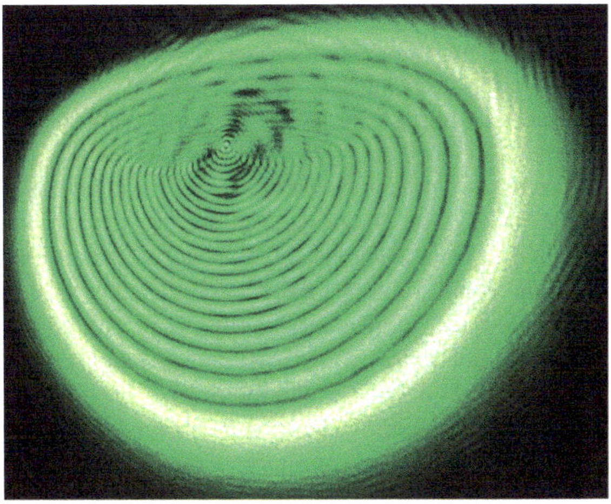

Fig. 3 Diffraction pattern of laser beam (532 nm) passing through the Au nanoparticles solution

light due to their surface plasmon resonance. This phenomenon produces high energetic (hot) electrons [29]. The excited energetic electrons quickly exchange their energy with other electrons or phonons and increase the temperature of the solution of gold nanoparticles around the path of laser beam. Initially the temperature of the solution decreases upward from the laser light beam, after some time heated solution across the light beam is replaced by above cool parts due to convection flow, causing above the laser light the motion of the solution increases. The lower portion of the solution from the laser light beam is not affected as that upper portion. The refractive index of the heated liquid reduces in the upper half part than the lower part of beam. As a result, the diffraction ring pattern of laser light beam becomes compressed from the top instead of nearly round, as shown in Fig. 3. Thus, surface plasmon excitation wavelength plays a major role in generating diffraction rings patterns [30, 31]. The ring pattern is an important phenomenon in the laser beam light corresponding to the surface plasmon resonance absorption band and can play an important role in the development of different optoelectronic and photonic devices.

2.2 Au-TiO_2 Nanocomposites

The TiO_2 surface consists of either oxygen vacancies of hydrogen absorption sites. In the former case, the number of oxygen atoms is less than that indicated by the Au-TiO_2 formula, these surface defects act as nucleation sites for Au deposition. Electronic changes have been observed in the presence of Au nanoparticles near Ti ions where the Ti 2*p* level underwent a band-bending of 0.15 eV [32, 33]. The bombardment of argon ion on TiO_2 led to surface defects such as oxygen vacancies and Ti^{3+} ions whereby deposition of Au on these defects led to a negative shift in the binding energy caused by a charge transfer from Au to Ti^{3+} ions. Oxygen vacancies may be more present near Au nanoparticles in consequence of the Schottky junction at Au-TiO_2 interface. The charge transfer is possible with Au particles big enough to exhibit high electronegativity. The placing a Au atom on top of a five-fold Ti atom after surface relaxation, led to a shift in the Au d-band closer to the Fermi energy and metal induced gap states were unlikely to be present [34, 35].

The surface plasmons are resonantly excited electromagnetic waves propagating along the Au-TiO_2 interface and are greatly localized in its region. An enhanced shift in the plasmon angles are observed upon coupling between the localized surface plasmon of the Au nanoparticles and the surface plasmon wave associated with the bulk dielectric. Small spherical clusters of gold display a narrow absorption peak at 520 nm which is associated with the surface plasmon resonance of gold's conduction electrons. When Au nanoparticles were imbedded in the Ti matrix, they also concluded that a shift in the absorption peak to a value around 600 nm was only observed when the annealing temperature of the film exceeds 300 °C and attributed this shift to the increase in the refractive index of the Ti matrix [13, 36, 37]. Furthermore, the particle morphology and concentration of Au nanoparticles did not affect the absorption spectra with samples annealed at the same temperature which contradicts observations made by other groups who stressed the fact that the plasmon resonance wavelength is strongly dependent on the metal nanoparticle size, shape, loading and the dielectric property of the surrounding medium.

The absorption of the plasmon by metal nanoparticles itself via internal electron–hole pair excitation, is called Landau damping. Landau damping depends on the particle size that is dominant for smaller particles size because the smaller particles have higher absorption cross-section than the scattering cross-section. This higher absorption creates excited (hot) electron–hole pairs in the metal nanoparticles. When the conduction band of TiO_2 and excited hot energy levels of metal nanoparticle are in contact with each other and proper matches then the generated hot electrons are transferred from metal to the conduction band of TiO_2. This charge transfer process at the interfacial contact between noble metal and TiO_2 creates a Schottky barrier (such as the Schottky barrier between Au and TiO_2 is about 1 eV) [38–40]. This barrier acts as trapping center for the electrons that inhibiting the electron–hole pair recombination, so the particle size of the metal should be closer to the mean free path of the electron. The particle size and preparation methods play an important role in enhances photocatalytic activity of metal-TiO_2 nanocomposites. If the particles size smaller than the mean free path of the electron then there is highly probability of electrons that these go through tunnel into TiO_2 semiconductor. In this process the surface plasmon decay, it is called chemical interface damping [41].

2.3 *Optical Nonlinearity*

The enhancement of local electric field in metal-TiO_2 nanocomposites is responsible for large enhancement in optical nonlinearity. The third order optical susceptibility of metal nanoparticles can be written as,

$$\chi^{(3)} = \frac{1}{3}\left(\frac{e^4}{K_F^4 E_F^3}\right)\left(\frac{2E_F}{E_g}\right)^6$$

where K_F is the Fermi wave vector, E_F is the Fermi energy and E_g is the energy band gap of material [42]. Hence the optical nonlinearity depends on the Fermi energy of the material. The accumulated charge carriers shift the Fermi energy level towards the conduction of the semiconductor. The shifting in Fermi energy level can pay a major role in enhances optical nonlinearity. The metal nanoparticles are excited by a laser beam of wavelength 532 nm, the wavelength close to surface plasmon resonance of the nanoparticles. This process generates highly energetic electrons near the surface of metal nanoparticle that enhance local electric field on the noble metal nanoparticle surface. The relation between local electric field (E) and applied field (E_0) can be expressed as $E = [3\varepsilon_m/(\varepsilon_m + 2\varepsilon_d)]E_0 = fE_0$, where ε_m is the dielectric constants of the metallic particle, ε_d is the surrounding medium and f is the local field factor. At surface plasmon resonance, the local field factor attains its maximum value. The optical nonlinearity of the material depends strongly on the local field factor or the dielectric constant of metal nanoparticles [43–45]. Therefore, the optical nonlinearity increases with increasing in surface plasmon resonance. So, we can say that the increase in surface plasmon resonance decreases the work function (work function $\varphi = \varepsilon_s + (E_C - E_F), = \chi_s + (E_c - E_F)$, where ε_s, E_C and E_F are the electron affinity, conduction band energy and Fermi energy of semiconductor, respectively). This shows the work function and nonlinear optical constants are inversely proportional to each other. We can say the position of Fermi level plays an important role in altering optical nonlinearity of the plasmonic materials.

3 Nonlinearity of Nanocomposites

The optical nonlinearity response of Au-TiO_2 nanocomposites is due to pure electronic transfer effects in Au nanoparticles. For observing this nonlinear response in nanocomposites, the intensity of the light beam needs to exceed above a threshold value, that intensity at which transmittance of Au-TiO_2 solution is independent of position and the concentration of Au nanoparticles [46]. Above threshold value, the optical nonlinearity will be increased and decrease transmittance as one moves towards the focus, as shown in Fig. 4, a characteristic of negative nonlinear refractive index. In Au-TiO_2 nanocomposites, the optical nonlinearity occurs due to the dielectric confinement effect, which is enhanced by the surface plasmon resonance. The dielectric function of Au nanoparticle shows maximum value at wavelength 532 nm (2.33 eV) and therefore, the surface plasmon resonance effect of Au nanoparticles, which is directly related to the metal dielectric function or dielectric constant. The optical nonlinearity increases with increasing dopant concentration of Au due to

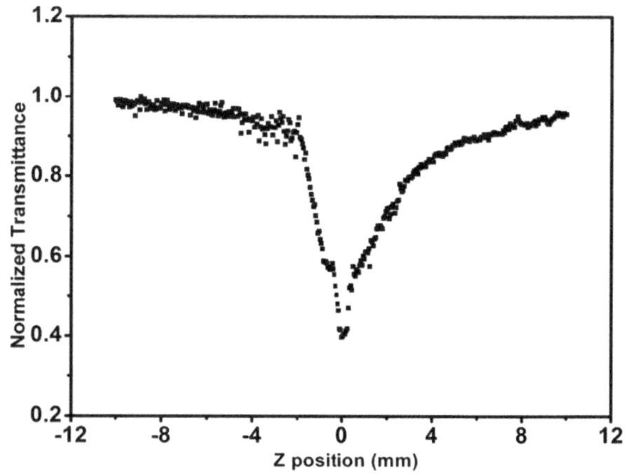

Fig. 4 Transmittance curve of Au-TiO_2 nanocomposite for concentration 7.5×10^{-3} of Au

the local electric field at wavelengths corresponding to the surface plasmon resonance absorption. Thus, the surface plasmon resonance helps in tuning the optical nonlinearity of the Au-TiO_2 nanocomposites [47–49]. Optical nonlinearity may find applications in nonlinear optics at wavelength 532 nm.

4 Conclusion

This chapter presents the interaction of laser beam of wavelength 532 nm with the Au nanoparticles suspension as well as Au-TiO_2 nanocomposites. The presence of Au nanoparticles provides a strong plasmonic absorption corresponding wavelength 522 nm. Laser beam of wavelength 532 nm is efficiently absorbed by the Au nanoparticles that change the refractive index and optical nonlinearity. The nonlinear optical response of materials (Au and Au-TiO_2 nanocomposites) at wavelength 532 nm shows the existence of an effective photon absorption behavior in nonlinearity. Thus, the surface plasmon resonance can play a major role in tuning the optical nonlinearity. The tunable nonlinear optical response can play a significant role in photovoltaic, solar cell, medical purposes, lasers, sensor, photovoltaic and other optical applications at wavelength 532 nm, a commonly available high-power laser.

5 Data Availability Statement

The data that support the findings of this study are available from the corresponding author upon reasonable request.

Acknowledgements Dr. Amodini Mishra acknowledges to ANRF-DST, Government of India, for providing financial support through the NPDF and women scientist (WISE-PDF) scheme.

Conflict of Interest The authors declare that they have no conflicts of interest.

References

1. Kumar, A., Mohanty, T.: Electro-optic modulation induced enhancement in photocatalytic activity of N-doped TiO_2 thin films. J. Phys. Chem. C **118**, 7130–7138 (2014)
2. Patel, A.S., Kumar, A., Mohanty, T.: Photo reduction altered work function of Au-TiO_2 nanoparticles measured by scanning Kelvin probe microscopy. J. Nanosci. Nanotechnol.Nanosci. Nanotechnol. **13**, 8217–8223 (2013)
3. Sundararaman, R., Narang, P., Jermyn, A.S., Goddard, W.A., Atwater, H.A.: Theoretical predictions for hot-carrier generation from surface plasmon decay. Nat. Commun.Commun. **5**, 5788–5795 (2014)
4. Mishra, A.: Effect of surface potential and charge transfer mechanism in reduced graphene oxide and magnetic nanocomposites. Mater. Res. Bull. **108**, 207 (2018)
5. Mishra, A., Mohanty, T.: One-step synthesis of Fe_3O_4/GO nanocomposites at 100 °C and its magnetic properties. Integr. Ferroelectr.. Ferroelectr. **184**, 178 (2017)
6. Karimzadeh, R.: Spatial self-phase modulation of a laser beam propagating through liquids with self-induced natural convection flow. J. Opt. **14**, 095701–095710 (2012)
7. Zhang, Y.-X., Wang, Y.-H.: Nonlinear optical properties of metal nanoparticles: a review. RSC Adv. **7**, 45129–45144 (2017)
8. Mishra, A., Kuanr, B.K., Mohanty, T.: Co_3O_4/reduced graphene oxide nanocomposite for removal of organic pollutants from aqueous medium. AIP Conf. Proc. **1832**, 050022 (2017)
9. Geints, Y.E., Panamarev, N.S., Zemlyanov, A.A.: Transient behavior of far-field diffraction patterns of a Gaussian laser beam due to the thermo-optical effect in metal nanocolloids. J. Opt. **13**, 055707–055715 (2011)
10. Mishra, A., Mishra, A., Pandey, A.: Study of organic pollutant removal capacity and work function of magnetite/graphene oxide nanocomposites. Mater. Res. Express **6**, 125039 (2019)
11. Wu, R., Zhang, Y., Yan, S., Bian, F., Wang, W., Bai, X., Lu, X., Zhao, J., Wang, E.: Purely coherent nonlinear optical response in solution dispersions of graphene sheets. Nano Lett. **11**, 5159–5164 (2011)
12. Patra, P.P., Chikkaraddy, R., Tripathi, R.P.N., Dasguptaand, A., Kumar, G.V.P.: Plasmofluidic single-molecule surface-enhanced Raman scattering from dynamic assembly of plasmonic nanoparticles. Nat. Commun.Commun. **5**, 4357–4364 (2014)
13. Boyd, R.W., Shi, Z., Leon, I.D.: The third-order nonlinear optical susceptibility of gold. Opt. Commun.Commun. **326**, 74–78 (2014)
14. Mishra, A., Mohanty, T.: Analysis of surface potential and magnetic properties of Fe_3O_4/ graphene oxide nanocomposites. AIP Conf. Proc. **1731**, 050016 (2016)
15. Meier, M., Wokaun, A.: Enhanced fields on large metal particles-dynamic depolarization. Opt. Lett. **8**, 581–583 (1983)
16. Clavero, C.: Plasmon-induced hot-electron generation at nanoparticle/metal-oxide interfaces for photovoltaic and photocatalytic devices. Nat. Photon. **8**, 95–103 (2014)
17. Mishra, A., Yadav, N., Bhattacharya, J., Moulick, R.G.: Covalent attachment of streptavidin to two-dimensional magnetic nanocomposite enhances surface-enhanced Raman spectroscopic signal. J. Appl. Phys. **125**, 165301 (2019)
18. He, K.X., Abeleldayem, H., Sekhar, P.C., Venkateswarlu, P., George, M.C.: Transient multiple diffraction rings induced by ultrafast laser from Chinese tea. Opt. Commun.Commun. **81**(1–2), 101–105 (1991)

19. Furube, A., Du, L., Hara, K., Katoh, R., Tachiya, M.: Ultrafast plasmon-induced electron transfer from gold nanodots into TiO_2 nanoparticles. J. Am. Chem. Soc. **129**, 14852–14853 (2007)
20. Kochuveedu, S.T., Kim, D.P., Kim, D.H.: Surface plasmon induced visible light photocatalytic activity of TiO_2 nanospheres decorated by Au nanoparticles with controlled configuration. J. Phys. Chem. C **116**, 2500–2506 (2012)
21. Mubeen, S., Lee, J., Singh, N., Kramer, S., Stucky, G.D., Moskovits, M.: An autonomous photosynthetic device in which all charge carriers derive from surface plasmons. Nat. Nanotechnol.Nanotechnol. **8**, 247–251 (2013)
22. Mishra, A., Mohanty, T., Kuanr, B.K.: Temperature-dependent magnetic behavior of α-Fe_2O_3/ GO nanocomposites. AIP Conf. Proc. **1942**, 050014 (2018)
23. Tian, Y., Tatsuma, T.: Mechanisms and applications of plasmon-induced charge separation at TiO_2 films loaded with gold nanoparticles. J. Am. Chem. Soc. **127**, 7632–7637 (2005)
24. Silva, C.G., Juarez, R., Marino, T., Molinari, R., Garcia, H.: Influence of excitation wavelength (UV or Visible Light) on the photocatalytic activity of titania containing gold nanoparticles for the generation of hydrogen or oxygen from water. J. Am. Chem. Soc. **133**, 595–602 (2011)
25. Srivastava, T., Shrivastav, A.M., Sinha, S., Polley, D., Alee, K.S., Sujatha, R.A., Mishra, A., Saxena, S.K.: Silicon nanowire: from fabrication to its application, in materials for electronic, magnetic, and spintronic technologies: characterization and applications from energy storage to disease detection, edited by A. Mishra (Springer Nature, Switzerland), pp. 41–63. (2024)
26. Kumar, A., Singh, R.P., Mohanty, T., Taneja, A.: Tunable optical nonlinearity of Au- TiO_2 nanocomposites. Photon. Nanostruct. Fundam. Appl. **33**, 1–9 (2019)
27. Hendrich, C., Bosbach, J., Stietz, F., Hubenthal, F., Vartanyan, T., Trager, F.: Chemical interface damping of surface plasmon excitation in metal nanoparticles: a study by persistent spectral hole burning. Appl. Phys. B **76**, 869–875 (2003)
28. Rodriguez, J.A., Lui, G., Jirsak, T., Hrbek, J., Chang, Z., Dvorak, J., Maiti, A.: Activation of gold on titania: adsorption and reaction of SO_2 on Au/TiO_2 (110). J. Am. Chem. Soc. **124**, 5242–5250 (2002)
29. Debeila, M.A., Raphulu, M.C., Mokoena, E., Avalos, M., Petranovskii, V., Coville, N., Scurrell, M.: The influence of gold on the optical properties of sol-gel derived Titania. Mat. Sci. Eng. A **396**, 70–76 (2005)
30. Chen, S., Murray, R.W.: Electrochemical quantized capacitance charg of surface ensembles of gold nanoparticles. J. Phys. Chem. B **103**, 9996–10000 (1999)
31. Kumar, A., Taneja, A., Mohanty, T., Singh, R.P.: Effect of laser beam propagation through the plasmonic nanoparticles suspension, Res. Opt. 3; 100081-100086 (2021). Wokaun, A.: Surface enhancement of optical fields, mechanism and applications, Mol. Phys. 56; 1-33 (1985)
32. Holland, W.R., Hall, D.G.: Surface-plasmon dispersion relation: shifts induced by the interaction with localized plasma resonances. Phys. Rev. B **27**, 7765–7768 (1983)
33. Lyon, L.A., Pena, D.J., Natan, M.J.: Surface plasmon resonance of Au colloid-modified Au films: particle size dependence. J. Phys. Chem. B **103**, 5826–5831 (1999)
34. Buso, D., Pacifico, J., Martucci, A., Mulvaney, P.: Gold-nanoparticle-doped TiO_2 semiconductor thin films: optical characterization, Adv. Funct. Mater. 17; 347-354 (2007). Miao, L., Shi, B., Yi, J., Jiang, Y., Zhao, C., Wen, S.: Ultrafast nonlinear optical response in solution dispersions of black phosphorus. Sci. Rep. 7; 3352-3359 (2017)
35. Wang, W., Wu, Y., Wu, Q., Hua, J., Zhao, J.: Coherent nonlinear optical response spatial self-phase modulation in $MoSe_2$ nano-sheets. Sci. Rep. **6**, 22072–22077 (2016)
36. Rahulan, K.M., Ganesan, S., Aruna, P.: Synthesis and optical limiting studies of Au-doped TiO_2 nanoparticles, Adv. Nat. Sci. Nanosci. Nanotechnol. 2; 025012-025017 (2011). Kumar, A., Kanjilal, D., Mohanty, T.: Evolution of damage fraction due to dense ionizing irradiation on TiO_2 film, Appl. Surf. Sci. 282; 595-600 (2013)
37. Prot, D., Stout, D.B., Lafait, J., Pincon, N., Palpant, B., Debrus, S.: Local electric field enhancements and large third-order optical nonlinearity in nanocomposite materials, J. Opt. A Pure Appl. Opt. 4; S99-S102 (2002). Dimitrov, V., Sakka, S.: Linear and nonlinear optical properties of simple oxides. II, J. Appl. Phys. 79; 1741–1745 (1996)

38. Chen, S.-Q., Liu, Z.-B., Zang, W.-P., Tian, J.-G., Zhou, W.-Y., Song, F., Zhang, C.-P.: Study on z-scan characteristics for a large nonlinear phase shift. J. Opt. Soc. Am. **22**, 1911–1916 (2005). https://doi.org/10.1364/JOSAB.22.001911
39. Kumari, S., Khare, A.: Studies on nonlinear response of epitaxial ruby thin film. IEEE J. Quantum Electron. **50**, 645–650 (2014)
40. Irimpan, L., Krishnan, B., Nampoori, V.P.N., Radhakrishnan, P.: Nonlinear optical characteristics of nanocomposites of ZnO-TiO_2-SiO_2. Opt. Mater. **31**, 361–365 (2008)
41. Geethakrishnan, T., Palanisamy, P.K.: Determination of the third-order optical nonlinearity of a triphenylmethane dye using 633 nm He-Ne laser. Opt. Commun.Commun. **270**, 424–428 (2007)
42. Mezher, M.H., Nady, A., Penny, R., Chong, W.Y., Zakaria, R.: Z-scan studies of the nonlinear optical properties of gold nanoparticles prepared by electron beam deposition. Appl. Opt. **54**, 9703–9708 (2015)
43. Hari, M., Joseph, S.A., Mathew, S., Radhakrishnan, P., Nampoori, V.P.N.: Band-gap tuning and nonlinear optical characterization of Ag:TiO_2 nanocomposites. J. Appl. Phys. **112**, 074307–074315 (2012)
44. Kumar, A., Patel, A.S., Mohanty, T.: Correlation of photodegradation efficiency with surface potential of silver-TiO_2 nanocomposite thin films. J. Phys. Chem. C **116**, 20404–20408 (2012)
45. Lee, H.W., Lee, K., Cho, S., Anthony, J.K., Lee, S., Lim, H., Kim, K., Pasiskevicius, V., Rotermund, F.: Evaluation of the third-order optical nonlinearity of Au:SiO_2 nanocomposites in the off-resonant spectral region. Opt. Commun.Commun. **286**, 347–352 (2013)
46. Conforti, M., Valle, G.D.: Derivation of third-order nonlinear susceptibility of thin metal films as a delayed optical response. Phys. Rev. B **85**, 245423–245428 (2012)
47. Brongersma, M.L., Halas, N.J., Nordlander, P.: Plasmon-induced hot carrier science and technology. Nat. Nanotechnol.Nanotechnol. **10**, 25–34 (2015)
48. Wu, Y., Wu, Q., Sun, F., Zhao, J.: Emergence of electron coherence and two-color all-optical switching in MoS2 based on Spatial Self-Phase Modulation. Proc. Natl. Acad. Sci. U.S.A. **112**(38), 11800–11805 (2015)
49. Kumar, A., Jaiswal, M.K., Kanjilal, D., Joshi, R.K., Mohanty, T.: Fermi level shifting of TiO_2 nanostructures during dense electronic excitation. Appl. Phys. Lett. **99**, 013109–013111 (2011)

Magnetic Nanoparticles and its Application in Oncological Disease Detection and Therapeutics

Neelu Singh, Monoj Kumar Das, Amodini Mishra, and Paulraj Rajamani

Abstract Increasing death toll globally due to cancer is the outcome of challenges not only in oncological disease detection but also in therapeutics. This led scientific community to explore the novel approach for cancer disease diagnosis and its treatment. Magnetic nanoparticles during the recent years in the field of cancer received considerable advancement, due to their certain properties which can be controlled by the various approaches of synthesis such as: hydrothermal, co-precipitation, thermal decomposition and polyol synthesis. However, significant increase in their biocompatibility can be achieved by the functionalization of NPs. Such tailored nanoparticles showed potential in enhanced imaging technology for cancer detection and therapies. Common core element, which is being used are silver, gold, etc. further usage of has been used as they harbour certain excellent qualities such as superparamagnetic and biodegradability. This chapter seeks to elaborate the unique properties of magnetic nanoparticles and their diagnostic and therapeutic implications in the field of cancer biology.

Keywords Magnetic nanoparticles · Cancer immunotherapy · Photo thermal ablation · Gene delivery · FG bead

N. Singh · P. Rajamani (✉)
School of Environmental Sciences, Jawaharlal Nehru University, New Delhi, India
e-mail: paulrajr@yahoo.com

M. K. Das
Department of Medicine, Division of Nephrology and Hypertension, Stony Brook University, Stony Brook, NY, USA

A. Mishra
Department of Physics, Indian Institute of Technology Delhi, New Delhi, India

A. Mishra and V. Pathak (eds.), *Smart Materials for Energy Storage and Biomedical Applications*, Engineering Materials, https://doi.org/10.1007/978-3-031-92584-9_4

1 Introduction

The current century evidenced the advent, growth and expansion of nanotechnology, which deals with the application of nanoparticles, size ranges from 1 to 100 nm at least in one dimension to the various fields such as chemistry, physics, biology thus, being produced at large level to satisfy its demand by various fields due to which it has strong socio-economic effect. Among the various disciplines these nanoparticles played a significant role such as health care systems where their potential applications are being explored from disease diagnosis to therapeutic approach. As per WHO cancer found to be leading cause of death globally, 2020 itself witnessed approx.10 million deaths due to cancer among which Lung, Colo rectal, liver, stomach and breast are the common cause of cancer death. To reduce the mortality early detection of cancer can lead to greater possibility to respond the treatment and thus increase the survival rate [1]. Among the various nanoparticles, magnetic nanoparticles (MNPs) exhibited certain remarkable property due to which they become one of the most promising materials for clinical diagnostic and therapeutic research and application [2].

These MNPs fabricated from Ni, Co, Fe, polymers etc [3]. Thus, these Magnetic NPs consisting of different metal or their oxides harbouring properties such as super-paramagnetic, high magnetic susceptibility, low Curie temperature, high coercivity etc. [4]. Due to these noble properties magnetic NPs are being explored for biomedical application like detection and bioseperation of biological objects (like nucleic acids, protein, bacteria, enzyme, etc.), targeted drug delivery [5]. thus, it is crucial to choose that which element should be used for the synthesis of nanoparticles with magnetic properties also whose chemical and physical properties can be attuned as per desired application, therefore these magnetic iron oxide NPs emerges as the toughest competitor harbouring all the adjustable critical properties which can be tailored further by compatible coating which can be later modified for the integration of various ligands resulting multifunctional MNPs [6]. Several types of Iron oxide nanomaterial have been investigated like Fe_3O_4, $FeIIFeIII_2O_4$, α-Fe_2O_3, γ-Fe_2O_3, FeO, ε-Fe_2O_3 and β-Fe_2O_3 [7] to determine the unique property that can be further explored.

2 Magnetic Nanoparticles Synthesis

There are various methods which are being used to synthesis magnetic nanomaterial such as physical, chemical and biological. ***Physical methods*** incorporate two approaches "Top-down" through which bulk materials are broken down to nano range, but has a limitation to achieve the desired shape and size, example Ball milling. Another approach is "bottom-up" through which synthesized nanoparticles are well dispersed of desired shape and size such as Laser Evaporation [4]. ***Chemical methods*** such as a.) **Coprecipitation** in which different metal ions such as $FeCl_3$ and $MnCl_2$ are coprecipitated in the presence of sodium hydroxide to form nanomaterial of

$MnFe_2O_4$, Manganese ferrite [8] here factors like metal ions, concentration and pH effect the size and shape [9], b.) **Thermal decomposition**, NPs produced at extreme temperature where organometallic precursors decomposed in presence of organic solvents, various stabilizers like oleic acid, hexadecylamine and fatty acids are used, and factors like precursor, type of surfactants, solvents, reaction time, and extent of temperature are adjusted to obtain to desired size and shape [10–12], c.) **Microemulsion** is a turbid system which primarily consist of hydrophilic and lipophilic phases involving surfactants. There are basically three types of microemulsions (1) W/O where oil is dominant with few droplets of water (2) O/W oil droplets in aquas phase (3) where both water and oil existed in equal amount [9], d.) **Hydrothermal synthesis** also known as **solvothermal** used to synthesis MNPs at high temperature and pressure during which hydrolysis and oxidation reaction takes place to synthesize MNPs. Formation of crystal largely depends on minerals solubility in water e.) **Sonochemical synthesis** here nanoparticles are synthesized due to chemical effects ascends from ultrasound resultant of acoustic cavitation [5, 13]. ***Biological synthesis*** is the known route of synthesis of MNPs using living organisms like bacteria, fungi, viruses etc. [14] such as *Actinobacter* sp. can synthesize maghemite under aerobic conditions synthesize also plant exudates [15], and extract from the different part of plants are being used to synthesis the MNPs with certain limitation like poor dispersion [5].

3 Stabilisation and Characterisation of Magnetic Nanoparticles

Thus synthesized MNPs are coated with various inorganic or organic material which further provide stabilisation and prevent loss of magnetic property. Inorganic coating includes silica, noble metals or metal oxides whereas Organic coating includes starch, dextran, PEG poly (ethylene glycol), PEI (polyethylenimine) etc. further functionalization of MNPs with several polymers having functional groups like (-NH_2, NH-, -COO- etc.) tend to respond accordingly as the pH changes, biomolecule like antibodies, enzymes, proteins, polypeptides, biotin etc. [16], a strategy to keep them remain stable till they reach target cellular site to bring the desired effect. Further characterisation of MNPs is performed using techniques like energy dispersed spectroscopy (EDS), high-resolution transmission electron microscopy (HR-TEM), X-ray diffraction (XRD), hydrodynamic size using dynamic light scattering (DLS) etc. [17], whereas magnetic properties can be determined by vibrating sample magnetometer or a SQUID-VSM magnetometer at a particular temperature and under the applied magnetic field [18] to assess the inherent characteristics of the material and its application for biomedical application [19].

4 Application of MNPs in Tumor Diagnosis

Among the various disease, world witnessing cancer as the leading cause of mortality and despite potential advancement to improve the disease diagnosis and therapeutics it poses healthcare challenge globally [20]. Therefore, along with the conventional therapy like radiation, chemo, and surgery to deal with the disease further novel advance techniques will be required that can easily detect pathological states in the early phases of oncogenesis, essential for improved diagnosis, treatment and survivalism. Because of the unique characteristics (low toxicity, magnetic property, easy surface functionalisation, biodegradability etc.) which Magnetic nanoparticles harbour gained the attention for application of cancer diagnosis and therapeutics [20].

4.1 Magnetic Nanoparticles in Cancer Detection and Therapeutics

MNPs in Imaging to detect any abnormalities, to monitor disease progression medical imaging is done [21] such as in magnetic resonance imaging (MRI), optical fluorescence imaging, positron emission tomography (PET), single-photon emission computed tomography (SPECT), computed tomography (CT), SERS etc.

Contrast and sensitivity of MRI can be enhanced by MNPs, due to their demonstrated potential, they are being used as contrasting agent and in therapeutics serving as drug delivery agent, gene delivery agent, for thermal ablation etc. [22].

Conventionally ELISA test was conducted from serum which uses antigen-antibody reaction for oncological detection monitoring cancer marker pre- and post-surgery and it is a time-consuming enzymatic reaction whereas, antibody-enclosed MNPs can be directed to immobilized antigen in presence of magnetic field [23] such as cancer biomarker can be detected using FG, where ferrite of 35–40 nm ferrite linked with adaptor molecule later layered with styrene copolymer followed by glycidyl methacrylate (GMA), to generate the FG to which specific ligands can bound later, can attracted to the target molecule and because of the ferrite core, under magnetic force can be easily separated or attracted [24].

In general, extracellular vesicles (EVs) are granular with size 50–150 nm are secreted by the cells consisting nucleic acids intracellularly, whose cell membrane are derived from proteins and lipids [25], thus derived EVs from cancer cells reflect surface protein like CD147 from colorectal cancer cells [26], CD91 from lung cancer [27] and HER2 from breast cancer cells [28] therefore, these specific EVs released from cancer cells can be detected using a unique assay known as ExoCounter system that utilizes FG bead to quantify EVs and explore the surface proteins at the same time. Here the whole mechanism is magnetically driven rapid sandwich immune assay using blue ray disc technology [29] FF (fluorescent FG beads) is a next generation FG beads where europium complexes that emit fluorescence at 618 nm encapsulated to FG beads, therefore in addition to prompt magnetic attraction their further signal amplification is not required, as it can be directly observed under fluorescence microscope this enables rapid disease diagnosis [30]. Another firm application of Magnetic NPs during intraoperative conditions that comprises SLN (sentinel lymph node) mapping and for the speedy diagnosis of metastasis in sentinel lymph node using a magnetic probe, where biocompatible MNPs like SPIONs (such as Resovist and Sienna+) are used, these SPIONs are 60 nm in size hydrophilic colloidal solution γ-Fe_2O_3 encapsulated with carboxydextran are readily taken up by the Kupffer cells not present in Tumorous tissue thus produces a contrasting effect in MRI [31]. Further magnetic probes are required to detect the SPIONs for SLN mapping in Sentinel Lymphatic nodes, such as Sentimag, MTJ sensor, DiffMag they all demonstrated the ability to detect SPIONS in 280–500 μg from 1 mm to 2.5 cm distance [32, 33].

Additionally, for pathological examination of Resected SLNs immunostaining system is one of the accurate diagnostic approach used for the lesion characteristics which is avidin–biotin complex four steps method where binding of antigen-antibody complex is time consuming which is its limitation for quick intraoperative diagnosis [34] therefore to shorten the time FF bead were used to establish MRIF (Magnetically Promoted Rapid Immunofluorescence) staining thus shortening the reaction time using magnet later after careful washing can be observed under fluorescence microscope directly [35].

Drug delivery: Due to their tunable property such as surface modification and size distribution, can be used for chemotherapeutics and biotherapeutics to check tumor growth by disrupting DNA replication, cell function, cell division, protein expression, or anti-apoptotic mechanism. Anticancer drug with premium anticancer efficacy, but their long-term usage can lead to severe toxicity like irreversible kidney and heart lesions and bone marrow suppression therefore to reduce the clinical toxicity thermos-chemotherapeutic strategy was used for liver cancer treatment such as MTRN/DOX (magnetothermally responsive nanocarrier/doxorubicin) was designed containing MZF-MNPs (Mn-Zn Ferrite-magnetic nanoparticles), copolymer drug carter (thermosensitive), and DOX (doxorubicin, anticancer drug). Thus, effective magnetic targeting achieved under the magnetic field concentrating the nanocarrier at the tumor spot allowing the controlled release of drug in the tumor site with prolonged retention time along with reduced distribution to the vital organs like heart and kidney thus reducing the side effects [36].

Gene delivery: Another major advantage of these nanoparticles is that they can be used for ***gene delivery*** to tumor cells such as MNPs conjugated with biomolecule like siRNA that interfere with the translation process in the cellular environment and thus inhibiting gene expression in cancer cells [37]. Kievit et al. developed nanovector to investigate targeted gene delivery to G6 glioma cells where at the core iron oxide nanoparticles coated with PEG and PEI. Further GFP encoding DNA was bound followed by CTX (chlorotoxin) a targeting ligand attached using a PEG linker. Administration of thus synthesized nano-vectors denoted as NP: DNA-CTX found to increase in target tumor tissue as compared to control receiving NP: DNA formulation with the minimal cellular uptake to other organs like kidney, spleen and liver [38].

Cancer immune therapy is another approach that incorporates patient's immune system to recognize cancer cells thus later check their growth. Therefore, various approaches are being used to elicit the response of immune cells one is cancer immunotherapy along with magnetic nanoparticles. Chiang et al demonstrated that fucoidan-dextran based magnetic nanoparticle coupled with anti-PD-L1 (checkpoint inhibitor) and anti-CD28/anti-CD3(T-cell activators), can repairs the immunosuppressive tumor microenvironment by reviving tumour-infiltrating lymphocytes, along with targeting the nanomedicine via magnetic steering to the tumour to further diminish the off-target effects [39]. Also, MNPs is being used as anticancer agent by various groups, a study reported that, ferumoxytol nanoparticles can not only effectively inhibit tumor growth but also prevented metastasis [40], apart from this magnetic nanoparticle are being used for Tumor ablation therapy that includes magnetic hyperthermia (tumor destruction from the heat produced by MNPs under magnetic field), photothermal therapy (cancer cell destruction from heat generated from MNPs upon light exposure), photodynamic therapy (cancer cell death induced by ROS generated from MNPs when exposed to light) [20]. In a study A33 antibody conjugated with gold shells-iron oxide core nanoparticles targeted for photothermal treatment against colorectal cancer and observed that cells with A33 antigen exhibited increased accumulation of NPs and killing when exposed to NIR radiation then to those non A33 antigen expressing cells [41] Table 1. Representing list of MNPs functionalization/modification for the cancer detection and therapeutics (Fig. 1).

Therefore, several scientific studies revealing unique magnetic properties along with surface functionalisation make them potential candidate which can be used further for cancer diagnosis and therapeutics, however concrete more study on MNPs are need to explore for their biomedical application.

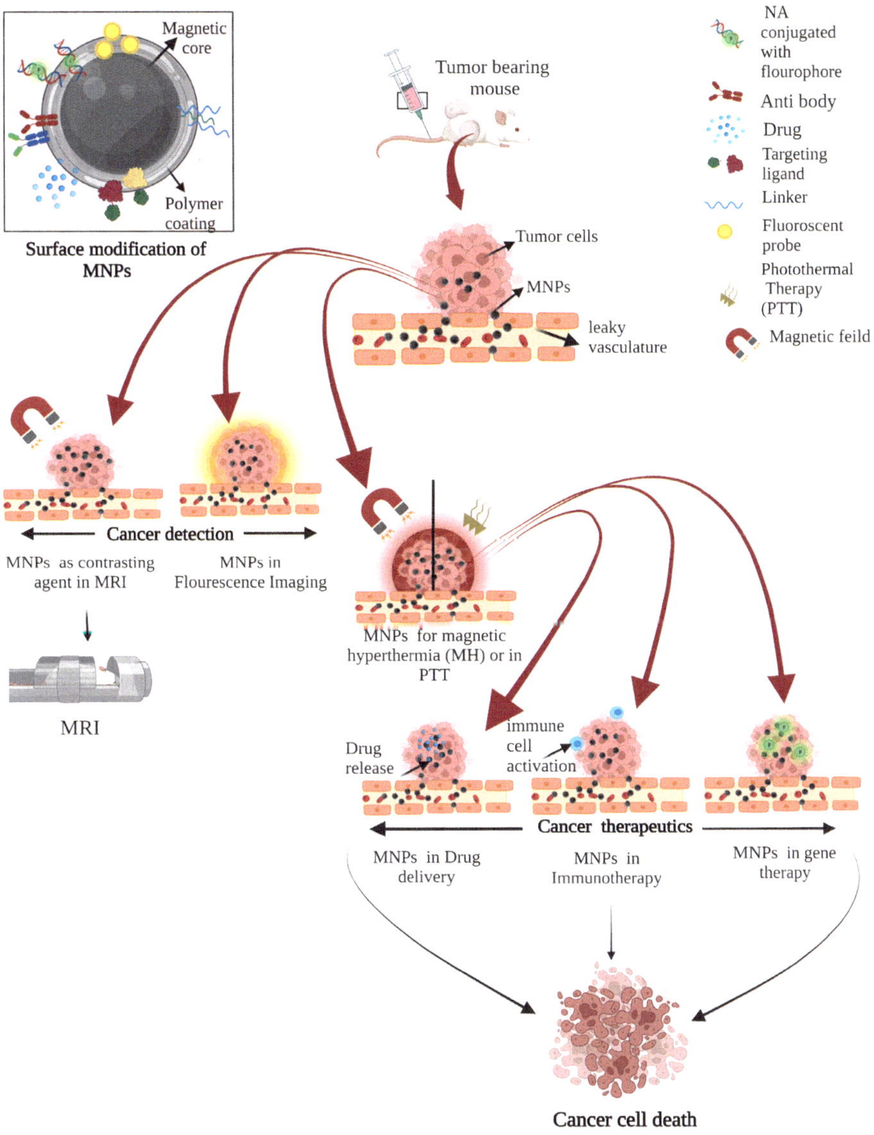

Fig. 1 Schematic representation of MNPs surface functionalised with various ligands and their role in cancer diagnosis and therapeutics

Table 1 List of MNPs functionalized/modified to oncological detection and therapeutics

S. No.	MNPs and its surface modification	Application	References
1	Fe3O4 modified with APTES further functionalized with heparin–pheophorbide-A conjugates	MR/Dual mode fluorescence Imaging and photodynamic therapy	[42]
2	Poly (DMA-r-mPEGMA-r-MA) hybrid Au–Fe3O4 nanoparticles	CT/MR imaging	[43]
3	Iron oxide nanoparticle loaded with doxorubicin (IONPs-DOX) further modified with PEG and LHRH	Targeted drug delivery and mild hyperthermia	[44]
4	Anti-HER2 antibody conjugated on silane-coated MNPs	Cancer diagnosis	[45]
5	Fe3O4@SiO2(FITC)/PEI-FA/Notch-1 shRNA	Magnetic guided gene delivery as well as for MRI in cancer diagnosis	[46]
6	MNP functionalized with folic acid	Targeted cancer therapy	[47]
7	Fe3O4-Ce6-FA nanoparticles	Photodynamic therapy (PDT)	[48]
8	Iron oxide nanoparticles functionalized with diosgenin (IONPs-D)	Cancer therapy	[49]
9	Fe3O4 magnetic nanoparticles modified with tyrosine loaded with tamoxifen (F@Tyr@TMX NPs)	Anticancer drug delivery	[50]
10	MNPs coated with L-aspartic acid further conjugated with paclitaxel (F-Asp-PTX NPs)	Anticancer drug delivery	[51]
11	MNPs coated with carbopol followed by Chitosan and was loaded with gemcitabine	Anticancer drug delivery	[52]
12	Superparamagnetic iron oxide nanoparticles coated with heparin-Poloxamer (HP) to form SPION@HP core–shell system loaded with Doxorubicin	Anticancer drug delivery	[53]
13	IOMNs coated with L-phenyl alanine loaded with chrysin (chrysin@Phe@IOMNs)	Anticancer drug delivery	[54]
14	HER2 antibody conjugated with magnetic nanoparticles coated with glycerol monooleate (GMO-MNPs) functionalised with DMSA (2, 3 meso mercapto succinic acid)	Targeted anticancer drug delivery	[55]
15	Fe3O4 filled into CNTs conjugated with silica coated CdTe quantum dots further coated with transferrin loaded with doxorubicin hydrochloride (DOX)	Magnetically guided drug delivery as well as cancer-targeted optical imaging	[56]

(continued)

Table 1 (continued)

S. No.	MNPs and its surface modification	Application	References
16	Surface functionalized Fe3O4	Photothermal therapy	[57]
17	Polydopamine (PDA)-coated magnetic nanoparticles functionalized with mono-6-thio-β-cyclodextrin (SH-βCD) loaded with doxorubicin (Fe3O4@PDA@SH-βCD@DOXO)	Magnetically guided drug delivery as well as MRI	[58]
18	MNPs loaded with curcumin (MNP-CUR)	Magnetically targeted drug delivery	[59]
19	Doxorubicin loaded onto CG-rich duplex containing prostate-specific membrane antigen (PSMA) aptamer-conjugated thermally cross-linked superparamagnetic iron oxide nanoparticles (TCL-SPIONs) (Dox@Apt–hybr–TCL-SPION)	Selective drug-delivery and tumor detection through MRI	[60]
20	Boron nitride nanotubes linked with Fe3O4 (BNNT-Fe3O4)	Magnetohyperthermia	[61]
21	Green fluorescent protein DNA-bearing polyethyleneimine-coated SPIONs	Gene delivery	[62]
22	Folic acid (FA)-targeted magnetic nanocomposite functionalized with tissue factor pathway inhibitor-2 (TFPI-2) and loaded with cisplatin (CDDP) (FA-MNP/CDDP/TFPI-2)	Magnetic targeted therapy and magnetic hyperthermia	[63]
23	AGO2 protein conjugated and anti-HER2 labelled Superparamagnetic iron oxide SP-AH further loaded with MIR376B combined with cisplatin	Targeted therapy	[64]
24	siRNA and metalloproteinase 10 (ADAM10) loaded into polyethylene glycol–polyethylenimine–functionalised Fe3O4 nanoparticles	Targeted therapy	[64]

References

1. https://www.who.int/news-room/fact-sheets/detail/cancer
2. Gul, S., Khan, S.B., Rehman, I.U., Khan, M.A., Khan, M.I.: A comprehensive review of magnetic nanomaterials modern day theranostics. Front. Mater. **6**, 179 (2019)
3. Alagiri, M., Muthamizhchelvan, C., Ponnusamy, S.: Structural and magnetic properties of iron, cobalt and nickel nanoparticles. Synth. Met. **161**(15–16), 1776–1780 (2011)
4. Ali, A., Shah, T., Ullah, R., Zhou, P., Guo, M., Ovais, M., Rui, Y.: Review on recent progress in magnetic nanoparticles: synthesis, characterization, and diverse applications. Front. Chem. **9**, 629054 (2021)
5. Wu, W., He, Q., Jiang, C.: Magnetic iron oxide nanoparticles: synthesis and surface functionalization strategies. Nanoscale Res. Lett. **3**(11), 397–415 (2008)

6. Farzin, A., Etesami, S.A., Quint, J., Memic, A., Tamayol, A.: Magnetic nanoparticles in cancer therapy and diagnosis. Adv. Healthc. Mater. **9**(9), 1901058 (2020)
7. Cornell, R.M., Schwertmann, U.: The iron oxides: structure, properties, reactions, occurrences, and uses, vol. 664. Wiley-vch, Weinheim (2003)
8. Chen, J.P., Sorensen, C.M., Klabunde, K.J., Hadjipanayis, G.C., Devlin, E., Kostikas, A.: Size-dependent magnetic properties of $MnFe_2O_4$ fine particles synthesized by coprecipitation. Phys. Rev. B **54**(13), 9288 (1996)
9. Mosayebi, J., Kiyasatfar, M., Laurent, S.: Synthesis, functionalization, and design of magnetic nanoparticles for theranostic applications. Adv. Healthc. Mater. **6**(23), 1700306 (2017)
10. Kudr, J., Haddad, Y., Richtera, L., Heger, Z., Cernak, M., Adam, V., Zitka, O.: Magnetic nanoparticles: from design and synthesis to real world applications. Nanomaterials **7**(9), 243 (2017)
11. Patsula, V., Kosinová, L., Lovrić, M., Ferhatovic Hamzić, L., Rabyk, M., Konefal, R., Horák, D.: Superparamagnetic Fe_3O_4 nanoparticles: synthesis by thermal decomposition of iron (III) glucuronate and application in magnetic resonance imaging. ACS Appl. Mater. Interfaces, 8(11), 7238-7247 (2016)
12. Lu, A.H., Salabas, E.E., Schüth, F.: Magnetic nanoparticles: synthesis, protection, functionalization, and application. Angew. Chem. Int. Ed. **46**(8), 1222–1244 (2007)
13. Suslick, K.S.: Sonochemistry. Science **247**(4949), 1439–1445 (1990)
14. Verma, R., Pathak, S., Srivastava, A.K., Prawer, S., Tomljenovic-Hanic, S.: ZnO nanomaterials: green synthesis, toxicity evaluation and new insights in biomedical applications. J. Alloy. Compd. **876**, 160175 (2021)
15. Bharde, A.A., Parikh, R.Y., Baidakova, M., Jouen, S., Hannoyer, B., Enoki, T., Sastry, M.: Bacteria-mediated precursor-dependent biosynthesis of superparamagnetic iron oxide and iron sulfide nanoparticles. Langmuir **24**(11), 5787–5794 (2008)
16. Jang, J.H., Lim, H.B.: Characterization and analytical application of surface modified magnetic nanoparticles. Microchem. J. **94**(2), 148–158 (2010)
17. Bhattacharjee, S.: DLS and zeta potential–what they are and what they are not? J. Control. Release **235**, 337–351 (2016)
18. Sivakumar, P., Ramesh, R., Ramanand, A., Ponnusamy, S., Muthamizhchelvan, C.: Synthesis and characterization of nickel ferrite magnetic nanoparticles. Mater. Res. Bull. **46**(12), 2208–2211 (2011)
19. Ferreira, M., Sousa, J., Pais, A., Vitorino, C.: The role of magnetic nanoparticles in cancer nanotheranostics. Materials **13**(2), 266 (2020)
20. Mukherjee, S., Liang, L., Veiseh, O.: Recent advancements of magnetic nanomaterials in cancer therapy. Pharmaceutics **12**(2), 147 (2020)
21. Massoud, T.F., Gambhir, S.S.: Molecular imaging in living subjects: seeing fundamental biological processes in a new light. Genes Dev. **17**(5), 545–580 (2003)
22. Vallabani, N.V., Singh, S.: Recent advances and future prospects of iron oxide nanoparticles in biomedicine and diagnostics. 3 Biotech **8**(6), 1–23 (2018)
23. Onishi, T., Mihara, K., Matsuda, S., Sakamoto, S., Kuwahata, A., Sekino, M., Kitagawa, Y.: Application of magnetic nanoparticles for rapid detection and in situ diagnosis in clinical oncology. Cancers **14**(2), 364 (2022)
24. Nishio, K., Masaike, Y., Ikeda, M., Narimatsu, H., Gokon, N., Tsubouchi, S., Handa, H.: Development of novel magnetic nano-carriers for high-performance affinity purification. Colloids Surf. B **64**(2), 162–169 (2008)
25. Belov, L., Matic, K.J., Hallal, S., Best, O.G., Mulligan, S.P., Christopherson, R.I.: Extensive surface protein profiles of extracellular vesicles from cancer cells may provide diagnostic signatures from blood samples. J. Extr. Vesicles **5**(1), 25355 (2016)
26. Millimaggi, D., Mari, M., D'Ascenzo, S., Carosa, E., Jannini, E.A., Zucker, S., Dolo, V.: Tumor vesicle—associated CD147 modulates the angiogenic capability of endothelial cells. Neoplasia **9**(4), 349–357 (2007)
27. Ueda, K., Ishikawa, N., Tatsuguchi, A., Saichi, N., Fujii, R., Nakagawa, H.: Antibody-coupled monolithic silica microtips for highthroughput molecular profiling of circulating exosomes. Sci. Rep. **4**(1), 1–9 (2014)

28. Marleau, A.M., Chen, C.S., Joyce, J.A., Tullis, R.H.: Exosome removal as a therapeutic adjuvant in cancer. J. Transl. Med. **10**(1), 1–12 (2012)
29. Yokose, T., Kabe, Y., Matsuda, A., Kitago, M., Matsuda, S., Hirai, M., Kitagawa, Y.: O-glycan-altered extracellular vesicles: a specific serum marker elevated in pancreatic cancer. Cancers **12**(9), 2469 (2020)
30. Sakamoto, S., Omagari, K., Kita, Y., Mochizuki, Y., Naito, Y., Kawata, S., Handa, H.: Magnetically promoted rapid immunoreactions using functionalized fluorescent magnetic beads: a proof of principle. Clin. Chem. **60**(4), 610–620 (2014)
31. Kato, N., Takahashi, M., Tsuji, T., Ihara, S., Brautigam, M., Miyazawa, T.: Dose-dependency and rate of decay of efficacy of Resovist on MR images in a rat cirrhotic liver model. Invest. Radiol. **34**(9), 551 (1999)
32. Houpeau, J.L., Chauvet, M.P., Guillemin, F., Bendavid-Athias, C., Charitansky, H., Kramar, A., Giard, S.: Sentinel lymph node identification using superparamagnetic iron oxide particles versus radioisotope: the French Sentimag feasibility trial. J. Surg. Oncol. **113**(5), 501–507 (2016)
33. Visscher, M., Waanders, S., Krooshoop, H.J., ten Haken, B.: Selective detection of magnetic nanoparticles in biomedical applications using differential magnetometry. J. Magn. Magn. Mater. **365**, 31–39 (2014)
34. Esbona, K., Li, Z., Wilke, L.G.: Intraoperative imprint cytology and frozen section pathology for margin assessment in breast conservation surgery: a systematic review. Ann. Surg. Oncol. **19**(10), 3236–3245 (2012)
35. Onishi, T., Matsuda, S., Nakamura, Y., Kuramoto, J., Tsuruma, A., Sakamoto, S., Kitagawa, Y.: Magnetically promoted rapid immunofluorescence staining for frozen tissue sections. J. Histochem. Cytochem. **67**(8), 575–587 (2019)
36. Li, M., Bu, W., Ren, J., Li, J., Deng, L., Gao, M., Wang, P.: Enhanced synergism of thermo-chemotherapy for liver cancer with magnetothermally responsive nanocarriers. Theranostics **8**(3), 693 (2018)
37. Yoo, D., Lee, J.H., Shin, T.H., Cheon, J.: Theranostic magnetic nanoparticles. Acc. Chem. Res. **44**(10), 863–874 (2011)
38. Kievit, F.M., Veiseh, O., Fang, C., Bhattarai, N., Lee, D., Ellenbogen, R.G., Zhang, M.: Chlorotoxin labeled magnetic nanovectors for targeted gene delivery to glioma. ACS Nano **4**(8), 4587–4594 (2010)
39. Chiang, C.S., Lin, Y.J., Lee, R., Lai, Y.H., Cheng, H.W., Hsieh, C.H., Chen, S.Y.: Combination of fucoidan-based magnetic nanoparticles and immunomodulators enhances tumour-localized immunotherapy. Nat. Nanotechnol. **13**(8), 746–754 (2018)
40. Zanganeh, S., Hutter, G., Spitler, R., Lenkov, O., Mahmoudi, M., Shaw, A., Daldrup-Link, E.: Iron oxide nanoparticles inhibit tumour growth by inducing pro-inflammatory macrophage polarization in tumour tissues. Nat. Nanotechnol. **11**(11), 986–994 (2016)
41. Kirui, D.K., Rey, D.A., Batt, C.A.: Gold hybrid nanoparticles for targeted phototherapy and cancer imaging. Nanotechnology **21**(10), 105105 (2010)
42. Nafiujjaman, M., Revuri, V., Nurunnabi, M., Cho, K.J., Lee, Y.K.: Photosensitizer conjugated iron oxide nanoparticles for simultaneous in vitro magneto-fluorescent imaging guided photodynamic therapy. Chem. Commun. **51**(26), 5687–5690 (2015)
43. Kim, D., Yu, M.K., Lee, T.S., Park, J.J., Jeong, Y.Y., Jon, S.: Amphiphilic polymer-coated hybrid nanoparticles as CT/MRI dual contrast agents. Nanotechnology **22**(15), 155101 (2011)
44. Taratula, O., Dani, R.K., Schumann, C., Xu, H., Wang, A., Song, H., Taratula, O.: Multifunctional nanomedicine platform for concurrent delivery of chemotherapeutic drugs and mild hyperthermia to ovarian cancer cells. Int. J. Pharm. **458**(1), 169–180 (2013)
45. Haghighi, A.H., Faghih, Z., Khorasani, M.T., Farjadian, F.: Antibody conjugated onto surface modified magnetic nanoparticles for separation of HER2+ breast cancer cells. J. Magn. Magn. Mater. **490**, 165479 (2019)
46. Yang, H., Li, Y., Li, T., Xu, M., Chen, Y., Wu, C., Liu, Y.: Multifunctional core/shell nanoparticles cross-linked polyetherimide-folic acid as efficient Notch-1 siRNA carrier for targeted killing of breast cancer. Sci. Rep. **4**(1), 1–10 (2014)

47. Niemirowicz, K., Car, H., Sadowska, A., Wątek, M., Kręowski, R., Cechowska-Pasko, M., Bucki, R.: Pharmacokinetics and anticancer activity of folic acid-functionalized magnetic nanoparticles. J. Biomed. Nanotechnol. **13**(6), 665–677 (2017)
48. Nam, K.C., Han, Y.S., Lee, J.M., Kim, S.C., Cho, G., Park, B.J.: Photo-functionalized magnetic nanoparticles as a nanocarrier of photodynamic anticancer agent for biomedical theragnostics. Cancers **12**(3), 571 (2020)
49. Ghosh, S., More, P., Derle, A., Kitture, R., Kale, T., Gorain, M., Chopade, A.: Diosgenin functionalized iron oxide nanoparticles as novel nanomaterial against breast cancer. J. Nanosci. Nanotechnol. **15**(12), 9464–9472 (2015)
50. Nosrati, H., Rashidi, N., Danafar, H., Manjili, H.K.: Anticancer activity of tamoxifen loaded tyrosine decorated biocompatible Fe_3O_4 magnetic nanoparticles against breast cancer cell lines. J. Inorg. Organomet. Polym. Mater. **28**(3), 1178–1186 (2018)
51. Tarantash, M., Nosrati, H., Kheiri Manjili, H., Baradar Khoshfetrat, A.: Preparation, characterization and in vitro anticancer activity of paclitaxel conjugated magnetic nanoparticles. Drug Dev. Ind. Pharm. **44**(11), 1895–1903 (2018)
52. Viota, J.L., Carazo, A., Munoz-Gamez, J.A., Rudzka, K., Gómez-Sotomayor, R., Ruiz-Extremera, A., Delgado, A.V.: Functionalized magnetic nanoparticles as vehicles for the delivery of the antitumor drug gemcitabine to tumor cells. Physicochemical in vitro evaluation. Mater. Sci. Eng. C **33**(3), 1183–1192 (2013)
53. Hoang Thi, T.T., Nguyen Tran, D.H., Bach, L.G., Vu-Quang, H., Nguyen, D.C., Park, K.D., Nguyen, D.H.: Functional magnetic core-shell system-based iron oxide nanoparticle coated with biocompatible copolymer for anticancer drug delivery. Pharmaceutics **11**(3), 120 (2019)
54. Nosrati, H., Javani, E., Salehiabar, M., Manjili, H.K., Davaran, S., Danafar, H.: Biocompatibility and anticancer activity of L-phenyl alanine-coated iron oxide magnetic nanoparticles as potential chrysin delivery system. J. Mater. Res. **33**(11), 1602–1611 (2018)
55. Dilnawaz, F., Singh, A., Mohanty, C., Sahoo, S.K.: Dual drug loaded superparamagnetic iron oxide nanoparticles for targeted cancer therapy. Biomaterials **31**(13), 3694–3706 (2010)
56. Chen, M.L., He, Y.J., Chen, X.W., Wang, J.H.: Quantum dots conjugated with Fe_3O_4-filled carbon nanotubes for cancer-targeted imaging and magnetically guided drug delivery. Langmuir **28**(47), 16469–16476 (2012)
57. Chu, M., Shao, Y., Peng, J., Dai, X., Li, H., Wu, Q., Shi, D.: Near-infrared laser light mediated cancer therapy by photothermal effect of Fe_3O_4 magnetic nanoparticles. Biomaterials **34**(16), 4078–4088 (2013)
58. Mrówczyński, R., Jędrzak, A., Szutkowski, K., Grześkowiak, B.F., Coy, E., Markiewicz, R., Jurga, S.: Cyclodextrin-based magnetic nanoparticles for cancer therapy. Nanomaterials **8**(3), 170 (2018)
59. Yallapu, M.M., Othman, S.F., Curtis, E.T., Bauer, N.A., Chauhan, N., Kumar, D., Chauhan, S.C.: Curcumin-loaded magnetic nanoparticles for breast cancer therapeutics and imaging applications. Int. J. Nanomed. **7**, 1761 (2012)
60. Yu, M.K., Kim, D., Lee, I.H., So, J.S., Jeong, Y.Y., Jon, S.: Image-guided prostate cancer therapy using aptamer-functionalized thermally cross-linked superparamagnetic iron oxide nanoparticles. Small **7**(15), 2241–2249 (2011)
61. Ferreira, T.H., Faria, J.A.Q.A., Gonzalez, I.J., Outon, L.E.F., Macedo, W.A.A., Gomes, D.A., Sousa, E.M.B.: BNNT/Fe_3O_4 system as an efficient tool for magnetohyperthermia therapy. J. Nanosci. Nanotechnol. **18**(10), 6746–6755 (2018)
62. Zuvin, M., Kuruoglu, E., Kaya, V.O., Unal, O., Kutlu, O., Yagci Acar, H., Koşar, A.: Magnetofection of green fluorescent protein encoding DNA-bearing polyethyleneimine-coated superparamagnetic iron oxide nanoparticles to human breast cancer cells. ACS Omega **4**(7), 12366–12374 (2019)
63. Liu, F., Chen, B., Chen, W., Chen, S., Ma, D., Xie, M.: Preparation of FA-targeted magnetic nanocomposites co-loading TFPI-2 plasmid and cis-platinum and its targeted therapy effects on nasopharyngeal carcinoma. Int. J. Med. Sci. **18**(11), 2355 (2021)
64. Panday, R., Abdalla, A.M.E., Yu, M., Li, X., Ouyang, C., Yang, G.: Functionally modified magnetic nanoparticles for effective siRNA delivery to prostate cancer cells in vitro. J. Biomater. Appl. **34**(7), 952–964 (2020)

A Theoretical Study on Two of the Bias-Based Methods of Optical Beam Defocus Measurement Involving Zernike Modes

Santanu Konwar, Nagendra Kumar, Karuna Sindhu Malik, and S. S. Goutam Buddha

Abstract This chapter explores two bias-based techniques for assessing defocus in an optical beam: the Modal Wavefront Sensor and the Astigmatic Defocus Sensor. Both methods rely on measuring intensity differences in the focal plane of a converging lens, providing a direct measure of defocus strength. These techniques are effective, simple, and easy to implement experimentally. This chapter presents the underlying principles, along with a detailed theoretical discussion and simulation results for both sensors.

Keywords Modal wavefront sensor · Astigmatic defocus sensor · Zernike polynomials · Displacement theorem · Sensitivity

1 Introduction

Defocus measurement is crucial across various fields due to its direct impact on visual clarity, image quality, and analytical accuracy. In areas such as microscopy [1], photography, and astronomy [2], precise focus is essential for capturing clear and detailed images. Accurate defocus measurement ensures that optical systems are properly calibrated, enabling researchers and professionals to obtain reliable data and make informed observations. This measurement is also vital in industries utilizing precision optics [3], in ophthalmology [4], in in-situ surface and thickness

S. Konwar (✉)
Department of Physics, Abhayapuri College, Bongaigaon, Assam, India
e-mail: k.santanu@abhayapuricollege.ac.in

N. Kumar
Department of Physics, Indian Institute of Technology Delhi, New Delhi, India

K. S. Malik
Department of Physics, Brainware University, Barasat, Kolkata, West Bengal, India

S. S. G. Buddha
Department of Physics, Tata Institute of Fundamental Research Hyderabad, Hyderabad, India

A. Mishra and V. Pathak (eds.), *Smart Materials for Energy Storage and Biomedical Applications*, Engineering Materials, https://doi.org/10.1007/978-3-031-92584-9_5

measurement in thin film deposition [5], in optometry [6], in optical beam trapping [7, 8] in digital cameras, smartphones, and other consumer electronics. Additionally, defocus measurement is fundamental in research fields requiring precise imaging and analysis, and it is increasingly important in emerging technologies like virtual reality (VR) and augmented reality (AR) [9]. Thus, by enhancing the technical performance of optical systems, defocus measurement improves diagnostic accuracy, research capabilities, and user experience across a wide range of applications. Its ongoing development and refinement drive innovation, enhances outcomes in scientific exploration and everyday visual technology use. Measuring defocus involves various techniques, each suited to different applications and levels of precision. Common methods include visual inspection, autofocus systems [10], depth from defocus measurement [11], the Foucault knife-edge test [12], interferometry [13, 14], phase retrieval methods [15], wavefront sensing [16–20], astigmatic lens approach [21–23] and optical coherence tomography (OCT) [24]. Each technique offers specific advantages and has its limitations, depending on factors such as the precision required, the application environment, and the type of optical system being analyzed. Often, combining multiple techniques can yield more comprehensive insights into defocus and other optical aberrations, ensuring accurate measurements and optimal performance of optical devices across diverse fields.

The current chapter will highlight two bias-based techniques of measuring defocus, i.e. the Modal wavefront Sensor [20, 25, 26] and the Astigmatic Defocus Sensor [21–23]. The Modal wavefront Sensor for defocus measurement works by dividing the incident beam into two beams of equal intensities, and a certain strength of defocus of opposite polarities is added to both beams. The strength of defocus is measured from the difference in intensities between the two beams focused on two different detectors. The Astigmatic Defocus Sensor, on the other hand, is based on the astigmatic lens approach where the incident beam is allowed to pass through an astigmatic lens and the strength of defocus is measured from the record of intensities of the beam focused on a detector. Both these techniques can directly measure defocus strength in an optical beam. Apart from these, they are very simple and easy to implement. This chapter presents the principle and an elaborate theoretical discussion with simulation results for both sensors.

2 Theoretical Backgrounds

For a better theoretical understanding of the two sensors, let us start with a brief theoretical background.

2.1 Calculation of the Focal Plane Intensity of a Lens

First, consider a monochromatic, collimated beam bearing a phase profile ψ. The beam is then allowed to focus by using a converging lens with a focal length of f (shown in Fig. 1). The amplitude of the complex field at a plane just after the lens L can be expressed as

$$A'_L(x, y) = A_L(x, y)e^{-j\frac{k}{2f}(x^2+y^2)} \quad (1)$$

where $A_L(x, y) = Ae^{j\psi(x,y)}$ is the amplitude of the complex field of the collimated beam incident on the lens, (x, y) are coordinates of the plane of the lens and k is a wave vector. The amplitude of the complex field at the back focal plane F of the lens can be calculated using the Fresnel diffraction integral [27]. Fresnel diffraction integral can be expressed as

$$A_F(f_x, f_y) = \frac{e^{jkf}}{j\lambda f}e^{j\frac{k}{2f}(f_x^2+f_y^2)}\int_{-\infty}^{\infty}\int_{-\infty}^{\infty}\left[A'_L(x, y)e^{j\frac{k}{2f}(x^2+y^2)}\right]e^{j\frac{2\pi}{\lambda f}(xf_x+yf_y)}dxdy \quad (2)$$

Here (f_x, f_y) are coordinates of the Fourier plane of the lens. Substituting the expression for $A_L\prime(x, y)$ from Eq. 1 and removing the constant phase e^{jkf}, we get

$$A_F(f_x, f_y) = \frac{e^{j\frac{k}{2f}(f_x{}^2+f_y{}^2)}}{j\lambda f}\int_{-\infty}^{\infty}\int_{-\infty}^{\infty}A_L(x, y)e^{-j\frac{2\pi}{\lambda f}((xf_x+yf_y)}dxdy \quad (3)$$

Polar coordinate systems are more convenient than those of the Cartesian coordinate system while representing the circular type of aperture. As most of the optical systems comprise a circular aperture, let us express the amplitude of the complex field in polar coordinates.

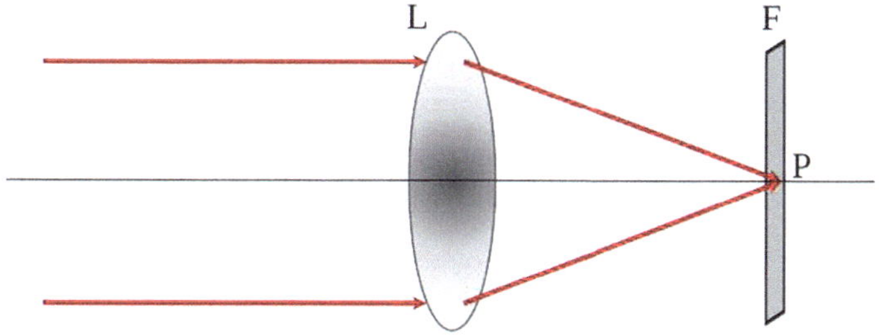

Fig. 1 A monochromatic collimated beam of light focused by lens L at the focal point P on the detector plane F

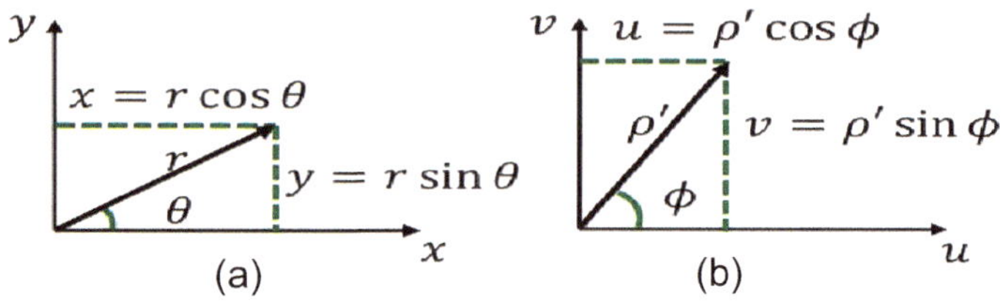

Fig. 2 Representation of (a) xy or $r\theta$ plane and (b) uv or $\rho\phi$ plane in Cartesian and Polar coordinates

The coordinates of the lens (x, y) in the Cartesian system can be converted to its polar coordinates (r, θ) as shown in Fig. 2a. Similarly, the back focal plane coordinates (u, v) of the lens in the Cartesian system can be converted into its polar coordinates (ρ', ϕ) as shown in Fig. 2b. Since the uv plane which is the back focal plane of the focusing lens lies on the Fourier plane ($f_x f_y$ plane) of the lens, the coordinates (ρ', ϕ) take the form $(\rho\lambda f, \phi)$, where ρ is the Fourier plane coordinate [27]. Thus, the coordinates (f_x, f_y) can be represented in polar coordinates (ρ, ϕ) as

$$\begin{aligned} f_x &= \rho\lambda f\cos\phi \Rightarrow \tfrac{f_x}{\lambda f} = \rho\cos\phi \\ f_y &= \rho\lambda f\sin\phi \Rightarrow \tfrac{f_y}{\lambda f} = \rho\sin\phi \end{aligned} \tag{4}$$

Now, the term $\frac{1}{\lambda f}(f_x x + f_y y)$ of Eq. 3 can be expressed in polar coordinates as

$$\tfrac{1}{\lambda f}(f_x x + f_y y) = \rho r(\cos\phi\cos\theta + \sin\phi\sin\theta) = \rho r\cos(\phi - \theta). \tag{5}$$

Thus, Eq. 3 in polar coordinates will take the form

$$A_F(\rho, \phi) = \frac{e^{j\frac{k}{2f}\rho^2}}{j\lambda f} \int_0^{2\pi}\int_0^{\infty} A_L(r, \theta)e^{-j2\pi\rho r\cos(\phi-\theta)} r dr d\theta \tag{6}$$

where $\rho^2\lambda^2 f^2 = f_x^2 + f_y^2$ and $A_L(r, \theta) = Ae^{j\psi(r,\theta)}$. The intensity at the focal point P of the lens can be obtained from the amplitude of the complex field as

$$\begin{aligned} I_F(\rho, \phi) = &\left[\frac{e^{j\frac{k}{2f}\rho^2\lambda^2 f^2}}{j\lambda f} \int_0^{2\pi}\int_0^{\infty} A_L(r, \theta)e^{-j2\pi\rho r\cos(\phi-\theta)} r dr d\theta\right] \\ &\times\left[\frac{e^{-j\frac{k}{2f}\rho^2\lambda^2 f^2}}{-j\lambda f} \int_0^{2\pi}\int_0^{\infty} A_L(r, \theta)e^{-j2\pi\rho r\cos(\phi-\theta)} r dr d\theta\right] \\ &= \frac{1}{\lambda^2 f^2}\left|\int_0^{2\pi}\int_0^{\infty} A_L(r, \theta)e^{-j2\pi\rho r\cos(\phi-\theta)} r dr d\theta\right|^2 \end{aligned} \tag{7}$$

2.2 Introduction to Zernike Polynomials

Optical aberrations can be accurately described using a series of polynomials known as Zernike polynomials. These polynomials form a complete and orthogonal set defined over a unit circle. First introduced by Fritz Zernike in 1934, they were originally used to assess mirror surface errors in his phase contrast technique. In this discussion, we will use Noll's convention for indexing the Zernike modes, one of several schemes available for the Zernike circle polynomials. According to Noll's convention, the Zernike modes are expressed in polar coordinates as follows. [14, 28],

$$\left.\begin{aligned} Z_{\text{even } k} &= \sqrt{n+1}R_n^m(r)\sqrt{2}\cos(m\theta) \\ Z_{\text{odd } k} &= \sqrt{n+1}R_n^m(r)\sqrt{2}\sin(m\theta) \end{aligned}\right\} \text{for } m \neq 0 \qquad Z_k = \sqrt{n+1}R_n^m(r) \;\Big\}\; \text{for } m = 0 \tag{8}$$

Here k denotes the index for the Zernike polynomials, while n represents the radial degree and m stands for the azimuthal frequency. Both n and m are nonnegative integers where m is less than or equal to n, and the difference between n and m must be even. The radial part of the Zernike polynomial, $R_n^m(r)$, can be written as,

$$R_n^m(r) = \sum_{q=0}^{\frac{(n-m)}{2}} \frac{(-1)^q(n-q)!}{q!\left[\frac{n+m}{2}-q\right]!\left[\frac{n-m}{2}-q\right]!} r^{n-2q} \tag{9}$$

Table 1 displays a few low-order Zernike polynomials [24].

An optical beam can exhibit none or several aberrations in its phase profile. Consequently, the phase profile of the optical beam, $\psi(r, \theta)$, can be expressed as a linear combination of these aberrations, with each aberration's strength indicated by a weight factor. Given that Zernike polynomials represent that optical aberration, the phase profile of an optical beam can be formulated as [19].

Table 1 Zernike polynomial representation of low-order aberrations

Index	(k) n m		Zk (r, θ)	Common aberration
4	2	0	$\sqrt{3(2r^2-1)}$	Defocus
5	2	2	$\sqrt{6(2r^2\sin 2\theta)}$	Astigmatism ± 45°
6	2	2	$\sqrt{6(2r^2\cos 2\theta)}$	Astigmatism 0°
7	3	1	$\sqrt{8(3r^2-2r)\sin\theta}$	y coma
8	3	1	$\sqrt{8(3r^2-2r)\cos\theta}$	x coma
9	3	3	$\sqrt{8r^2\sin 3\theta}$	Trefoil y
10	3	3	$\sqrt{8r^2\cos 3\theta}$	Trefoil x
11	4	0	$\sqrt{5(6r^4-6r^2+1)}$	Spherical aberration

$$\psi(r,\theta) = \sum_{k=0}^{\infty} a_k Z_k(r,\theta) \tag{10}$$

where a_k are coefficients that indicate the intensity of a specific Zernike polynomial $Z_k(r,\theta)$ within the optical beam's phase profile.

2.3 The Displacement Theorem

The displacement theorem allows the calculation of the shift in the focal plane based on the known root mean square (RMS) amplitude of defocus in the optical beam's phase profile. According to the displacement theorem [29], if H, K, L, and M are constants of the order λ, r is the radial coordinate, and Φ and Φ' are two aberration functions related in such a way that

$$\Phi = \Phi' + Hr^2 + Krsin\theta + Lrcos\theta + M \tag{11}$$

In that case, the three-dimensional intensity distribution near the focus remains unchanged except for a shift of the entire distribution according to the transformation

$$z' = z + 2\left(\frac{R}{w}\right)^2 H, x' = x + \left(\frac{R}{w}\right)K, y' = y + \left(\frac{R}{w}\right)L \tag{12}$$

where R represents the radius of the Gaussian reference sphere, and w is the radius of the optical system's exit pupil. The variables x, x', y, and y' are along the transverse directions, while z and z' are aligned with the optical axis. If Φ includes only an additional curvature and no tip or tilt, i.e., $\Phi = \Phi' + Hr^2$, then the three-dimensional intensity distribution will only experience a shift along the z direction by $\left[2\left(\frac{R}{w}\right)^2 H\right]$. This extra curvature is equivalent to the defocus aberration represented by the Zernike polynomial Z_4 as listed in Table 1. Therefore, the aberration function Φ can be written as $\Phi = \Phi' + \sqrt{3}\left(2r^2 - 1\right)$, leading to $H = 2\sqrt{3}$ when the constant term is ignored. Consequently, a 1 rad RMS amplitude of defocus aberration will cause a shift in the focal plane of $\frac{\lambda}{2\pi}\left[4\sqrt{3}\left(\frac{f}{w}\right)^2\right]$, where f is the optical system's focal length [27].

Now, let us move to the detailed discussion of the two types of bias-based techniques for measuring defocus.

3 Modal Wave Front Sensor

Mark A. Neil introduced the modal wave front sensor in 2000 [20], which extends the curvature sensor's capabilities to measure all aberration modes in an optical beam's phase profile. In this chapter, we will focus specifically on measuring defocus aberration. As depicted in Fig. 3, the sensor's principle involves an incident beam with a wave front W hitting a 50% beam splitter BS, splitting the beam into two equal-intensity beams. The defocus aberration of one of the beams is increased using a positive bias plate B_1, while the defocus aberration of the other beam is decreased by the same amount with a negative bias plate B_2. These beams are then directed to separate photodetectors D_1 and D_2 through lenses L_1 and L_2. The detectors are point detectors due to their small aperture. A mirror M aligns both the beams to be parallel to each other.

The difference in the intensities between the two detectors gives us the measure of the amplitude of defocus present in the beam. An elaborate theoretical discussion of the above principle is presented below.

If a beam with a RMS amplitude of Defocus (Z_4) in its phase profile passes through a focusing lens i.e. the lens L of Fig. 1, the wave front immediately after the lens will be.

$$W(r,\theta) = e^{jaZ_4(r,\theta)} \tag{13}$$

where $Z_4(r,\theta)$ denotes the Zernike mode representing defocus. The intensity of the focused beam at the lens's focal point can be represented as

$$I(\rho,\phi) = \frac{1}{\lambda^2 f^2} \times \left| \int_0^{2\pi} \int_0^{\infty} e^{jaZ_4(r,\theta)} e^{-j2\pi\rho r\cos(\phi-\theta)} r dr d\theta \right|^2 \tag{14}$$

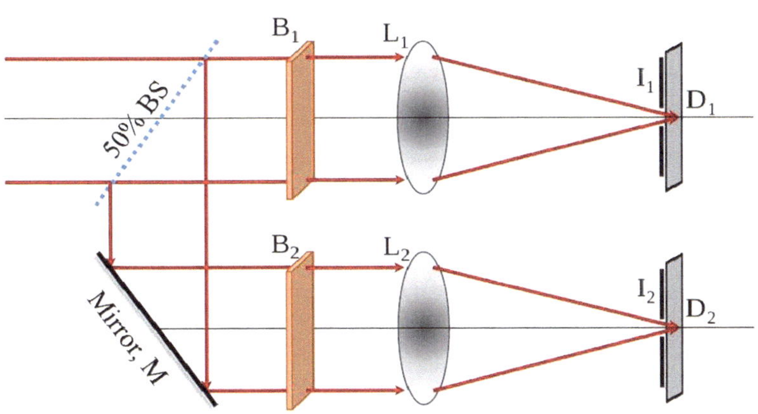

Fig. 3 Principle of modal wavefront sensor

For a detector with a finite radius $R = \rho_p \lambda f$, where ρ_p represents the Fourier coordinate of R, the intensity measured by this detector is obtained by integrating over the detector's area as follows

$$I_T(\rho, \phi) = \int_0^{2\pi} \int_0^{\rho_p} I(\rho, \phi) \rho d\rho d\phi \tag{15}$$

Now, incorporating the positive and the negative bias of $\pm b$ RMS amplitude of defocus (i.e. Zernike mode Z_4) to the two beams we get

$$I_{-b}(\rho, \phi) = \frac{1}{\lambda^2 f^2} \left| \int_0^{2\pi} \int_0^{\infty} e^{j(a-b)Z_4(r,\theta)} e^{-j2\pi\rho r \cos(\phi-\theta)} r dr d\theta \right|^2 \tag{16}$$

and

$$I_{+b}(\rho, \phi) = \frac{1}{\lambda^2 f^2} \left| \int_0^{2\pi} \int_0^{\infty} e^{j(a+b)Z_4(r,\theta)} e^{-j2\pi\rho r \cos(\phi-\theta)} r dr d\theta \right|^2 \tag{17}$$

Therefore, the intensities at the two detectors will be

$$I_1(\rho, \phi) = \int_0^{2\pi} \int_0^{\rho_p} I_{-b}(\rho, \phi) \rho d\rho d\phi \tag{18}$$

$$I_2(\rho, \phi) = \int_0^{2\pi} \int_0^{\rho_p} I_{+b}(\rho, \phi) \rho d\rho d\phi \tag{19}$$

The sensor output can now be obtained from the normalized intensity difference between the two detectors and can be expressed as [25, 26]

$$O_M = \frac{(I_1 - I_2)}{(I_1 + I_2)} \tag{20}$$

The root mean square (RMS) amplitude of defocus in the optical beam's phase profile can be determined by dividing the sensor's output by its sensitivity, given by $A_{RMS} = \frac{O_M}{S_M}$. In this equation, A_{RMS} represents the measured RMS amplitude of defocus, O_M denotes the output of the modal wavefront sensor, and S_M is the sensor's sensitivity to defocus measurement.

3.1 Sensitivity of the Sensor

The sensitivity of a sensor is determined by evaluating the gradient of the sensor output O_M with respect to the input aberration strength a, at $a = 0$, i.e. $\frac{\partial O_M}{\partial a}\big|_{a=0}$ [28]. Thus, we begin by calculating $\frac{\partial O_M}{\partial a}$ as follows

$$\begin{aligned}\frac{\partial O_M}{\partial a} &= \frac{(I_1+I_2)\frac{\partial}{\partial a}(I_1-I_2)-(I_1-I_2)\frac{\partial}{\partial a}(I_1+I_2)}{(I_1+I_2)^2}\\ &= \frac{2I_2\frac{\partial}{\partial a}(I_1)-2I_1\frac{\partial}{\partial a}(I_2)}{(I_1+I_2)^2}\end{aligned} \tag{21}$$

Now, applying the limit $a = 0$, we get

$$\frac{\partial O_M}{\partial a}\bigg|_{a=0} = \frac{2I_2\frac{\partial}{\partial a}I_1\big|_{a=0} - 2I_1\frac{\partial}{\partial a}I_2\big|_{a=0}}{(I_1+I_2)^2\big|_{a=0}} \tag{22}$$

While the modal wavefront sensor can measure various aberration modes, this paper focuses solely on the defocus aberration mode. Therefore, both the input and bias aberration modes are considered as defocus (i.e. Z_4) with a and b representing the RMS amplitudes of the input and bias aberration modes, respectively. Here, the expression for $\frac{\partial}{\partial a}\{I_{\pm b}(\rho, \phi)\}$ can be obtained as

$$\begin{aligned}\frac{\partial}{\partial a}\{I_{\pm b}(\rho,\phi)\} &= \frac{1}{\lambda^2 f^2}\frac{\partial}{\partial a}\left|\int_0^{2\pi}\int_0^1 e^{j(a\pm b)Z_4(r,\theta)}e^{-j2\pi\rho r\cos(\phi-\theta)}rdrd\theta\right|^2\\
&= \frac{1}{\lambda^2 f^2}\frac{\partial}{\partial a}\left[\int_0^{2\pi}\int_0^1 e^{j[(a\pm b)Z_4(r,\theta)-2\pi\rho r\cos(\phi-\theta)]}rdrd\theta\right.\\
&\quad\left.\times\int_0^{2\pi}\int_0^1 e^{-j[(a\pm b)Z_4(r,\theta)-2\pi\rho r\cos(\phi-\theta)]}rdrd\theta\right]\\
&= \frac{-j}{\lambda^2 f^2}\left[\int_0^{2\pi}\int_0^1 e^{j[(a\pm b)Z_4(r,\theta)-2\pi\rho r\cos(\phi-\theta)]}rdrd\theta\right.\\
&\quad\times\int_0^{2\pi}\int_0^1 Z_4(r,\theta)e^{-j[(a\pm b)Z_4(r,\theta)-2\pi\rho r\cos(\phi-\theta)]}rdrd\theta\\
&\quad-\int_0^{2\pi}\int_0^1 e^{-j[(a\pm b)}e^{Z_4(r,\theta)-2\pi\rho r\cos(\phi-\theta)]}rdrd\theta\end{aligned}$$

$$\times \int_0^{2\pi}\int_0^1 Z_4(r,\theta)e^{j[(a\pm b)Z_4(r,\theta)\,-2\pi\rho\cos(\phi-\theta)]}rdrd\theta \tag{23}$$

which results in

$$\begin{aligned}\frac{\partial}{\partial a}\{I_{\pm b}(\rho,\phi)\}\Big|_{a=0} &= \frac{-j}{\lambda^2 f^2}\Bigg[\int_0^{2\pi}\int_0^1 e^{j[\pm bZ_4(r,\theta)-2\pi\rho\cos(\phi-\theta)]}rdrd\theta \\ &\times \int_0^{2\pi}\int_0^1 Z_4(r,\theta)e^{-j[\pm bZ_4(r,\theta)-2\pi\rho\cos(\phi-\theta)]}rdrd\theta \\ &- \int_0^{2\pi}\int_0^1 e^{-j[\pm bZ_4(r,\theta)-2\pi\rho\cos(\phi-\theta)]}rdrd\theta \\ &\times \int_0^{2\pi}\int_0^1 Z_4(r,\theta)e^{j[\pm bZ_4(r,\theta)\,-2\pi\rho\cos(\phi-\theta)]}rdrd\theta\end{aligned} \tag{24}$$

If the incident beam is free from defocus, i.e. $a = 0$, the intensities at both the detectors will be equal, i.e. $I_1 = I_2$. As a increases, the intensity at one detector will rise while the intensity at the other will decrease. Consequently, the detectors will exhibit the opposite gradients with respect to a, such that $\frac{\partial}{\partial a}(I_1) = -\frac{\partial}{\partial a}(I_2)$. Applying these conditions to Eq. 22 yields

$$\frac{\partial O_M}{\partial a}\Big|_{a=0} = \frac{\frac{\partial}{\partial a}I_1\big|_{a=0}}{I_1\big|_{a=0}} \tag{25}$$

Thus, Eq. 25 gives the sensitivity of the modal wavefront sensor. By applying Eqs. 18, 24 and 25, the sensor's sensitivity can be computed for various RMS amplitudes (b) of the bias mode Z_4 and for different detector radii in Fourier plane coordinates (ρ_p).

Figure 4(a) illustrates the sensitivity of the sensor as a function of the RMS amplitude of the bias aberration mode Z_4 (i.e. b) and the detector radius in Fourier plane coordinates (ρ_p). In the plot, b is shown on the horizontal axis and ρ_p on the vertical axis, with sensitivity in a color range from − 0.2 to + 1.0. Figure 4(c) provides an enlarged view of Fig. 4(a) focusing on b ranging from 0 to 1.8 and ρ_p from 0.01 to 0.29. It is noted that for RMS amplitudes of Z_4 between $a = 1.2 - 1.4$ radian and $\rho_p \leq 0.01$, sensitivity exceeds 0.8. A sensitivity of approximately 0.9 is achieved at $\rho_p = 0.01$. Figure 4(b) presents a sensitivity plot versus b for $\rho_p = 0.01$, showing that the maximum sensitivity of 0.8942 occurs at $b = 1.3$ radian.

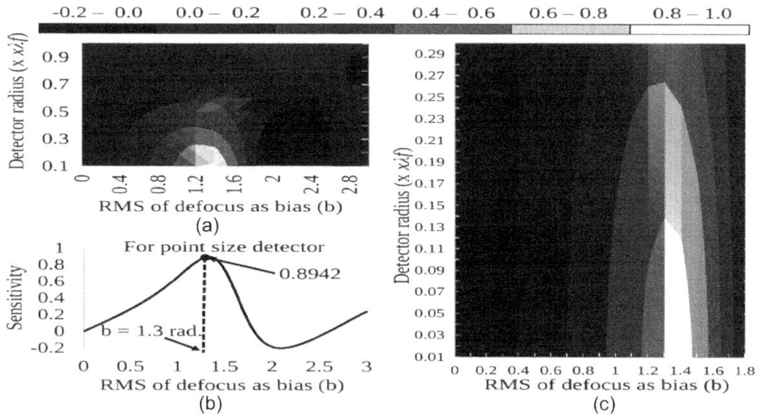

Fig. 4 Plot of the computed sensitivity of the modal wavefront sensor against (a) RMS amplitude of defocus (i.e. bias b) in radian and radius of the detector in Fourier coordinate (i.e. ρ_p), (c) RMS amplitude of defocus (i.e. bias b) in radian at $\rho_p = 0.01$ and (b) enlarged version of **a** for $b = 0-1.8$ and $\rho_p = 0.01 - 0.29$

3.2 *Cross-Sensitivity of the Sensor*

The cross-sensitivity of the wavefront sensor can be obtained using Eq. 22. However, the expression $I_{\pm b}(\rho, \phi)$ and $\frac{\partial}{\partial a}\{I_{\pm b}(\rho, \phi)\}\big|_{a=0}$ will get modified as

$$I_{\pm b}(\rho, \phi) = \frac{1}{\lambda^2 f^2}\left|\int_0^{2\pi}\int_0^{1} e^{jcZ_i(r,\theta)\pm jbZ_4(r,\theta)} e^{-j2\pi\rho r\cos(\phi-\theta)} r dr d\theta\right|^2 \tag{26}$$

where $Z_i(r, \theta)$ is the Zernike mode representing an aberration mode apart from the Defocus aberration mode due to the presence of which cross-sensitivity occurs and c is the RMS amplitude of the Z_i. The expression for cross-sensitivity can be written as

$$\begin{aligned}\frac{\partial}{\partial c}\{I_{\pm b}(\rho, \phi)\}\bigg|_{c=0} = \frac{-j}{\lambda^2 f^2}\Bigg[\int_0^{2\pi}\int_0^{1} e^{j[\pm bZ_4(r,\theta)-2\pi\rho r\cos(\phi-\theta)} r dr d\theta \\ \times \int_0^{2\pi}\int_0^{1} Z_i(r, \theta) e^{-j[\pm bZ_4(r,\theta)-2\pi\rho r\cos(\phi-\theta)} r dr d\theta \\ - \int_0^{2\pi}\int_0^{1} e^{-j[\pm bZ_4(r,\theta)-2\pi\rho r\cos(\phi-\theta)} r dr d\theta\end{aligned}$$

Table 2 Cross-sensitivity in a modal wavefront sensor, which is designed to measure defocus aberration mode, caused by various Zernike modes

Zernike modes	Z_6	Z_7	Z_8	Z_9	Z_{10}	Z_{11}
Cross-sensitivity	0	0	0	0	0	– 0.227

$$\times \int_0^{2\pi} \int_0^1 Z_i(r,\theta) e^{j[\pm bZ_4(r,\theta) - 2\pi\rho r\cos(\phi-\theta)} r dr d\theta \tag{27}$$

Table 2 displays the cross-sensitivities associated with the Zernike modes Z_6, Z_7, Z_8, Z_9, Z_{10} and Z_{11}.

Table 2 shows that the sensitivities for all other modes (Z_6, Z_7, Z_8, Z_9 and Z_{10}) are zero. However, for spherical aberration (Z_{11}), the sensitivity is – 0.227, which is approximately 58% lower than the sensitivity for defocus aberration.

4 Astigmatic Defocus Sensor

The Astigmatic Defocus sensor utilizes the astigmatic lens technique, which depends on the astigmatic deformation of an optical beam's phase profile. In this sensor, the optical beam initially passes through an astigmatic lens. Assume that the astigmatic lens is rotated by 45°. When a beam with a plane wavefront encounters such a lens, i.e. replacing lens L in Fig. 1 with an astigmatic lens, the wavefront immediately after the lens can be described as follows

$$W(r,\theta) = Ae^{jbZ_5(r,\theta)} \tag{28}$$

where $Z_5(r,\theta)$ denotes the Zernike mode representing astigmatism with a 45° tilt, serving as the bias aberration, and b is the RMS amplitude of this bias aberration (i.e. $Z_5(r,\theta)$). The intensity at the detector will now be

$$I(\rho,\phi) = \frac{1}{\lambda^2 f^2} \left| \int_0^{2\pi} \int_0^{\infty} e^{jbZ_5(r,\theta)} e^{-j2\pi\rho r\cos(\phi-\theta)} r dr d\theta \right|^2 \tag{29}$$

For a finite-sized detector with radius ρ_p, the total intensity across the detector is calculated by integrating over the entire detector area, i.e.

$$I_T(\rho,\phi) = \int_0^{2\pi} \int_0^{\rho_p} I(\rho,\phi) \rho d\rho d\phi \tag{30}$$

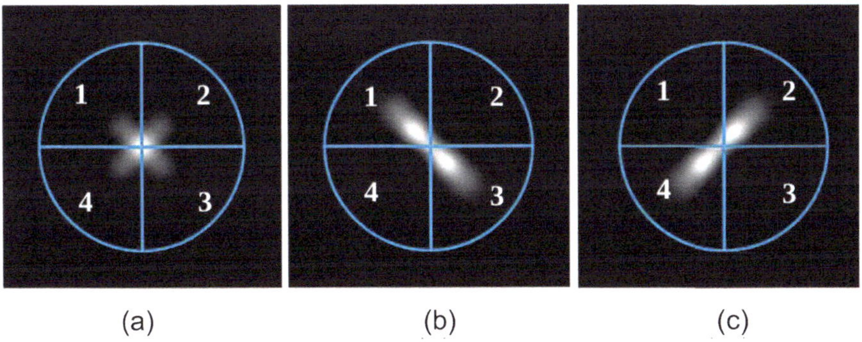

Fig. 5 The detector plane's intensity distribution with $b = 1$ radian and (a) $a = 0$ radian, (b) $a = 1$ radian and (c) $a = -1$ radian

If the detector aperture is divided into four equal sections as depicted in Fig. 5, the intensities of each sub-aperture will be

$$I_{1,2,3,4}(\rho, \phi) = \int_{\theta_1}^{\theta_2} \int_{0}^{\rho_p} I(\rho, \phi)\rho d\rho d\phi \tag{31}$$

Here $\theta_1 = 0$ & $\theta_2 = \frac{\pi}{2}$ for detector 1, $\theta_1 = \frac{\pi}{2}$ & $\theta_2 = \pi$ for detector 2, $\theta_1 = \pi$ & $\theta_2 = \frac{3\pi}{2}$ for detector 3 and $\theta_1 = \frac{3\pi}{2}$ & $\theta_2 = 2\pi$ for detector 4. Consider the incident beam to comprise a certain amount of defocus (e.g. b radian) represented by Z_4, as illustrated in Fig. 5(b, c). The figure shows that the presence of defocus introduces a specific directionality to the intensity pattern, and the opposite polarity of defocus alters the orientation of this pattern. Consequently, the intensity at the detector plane becomes

$$I(\rho, \phi) = \frac{1}{\lambda^2 f^2} \left| \int_0^{2\pi} \int_0^1 e^{jbZ_5(r,\theta)+jaZ_4(r,\theta)} e^{-j2\pi\rho r\cos(\phi-\theta)} r dr d\theta \right|^2 \tag{32}$$

The amount of defocus in the incident beam can be determined by recording the intensities across the different sub-apertures. The output of the sensor can then be calculated using the following expression [7]

$$O_A = \frac{(I_1+I_3)-(I_2+I_4)}{(I_1+I_3)+(I_2+I_4)} \tag{33}$$

It is evident from Fig. 5(b, c) that there is symmetry between diagonally opposite sub-apertures, with the intensities of these diagonally opposite subapertures being identical. Thus, $I_1 = I_3$ and $I_2 = I_4$ for both cases of $a = 1$ and $a = -1$. Consequently, the expression for the sensor output can be simplified to

$$O_A = \frac{(I_1 - I_2)}{(I_1 + I_2)} \tag{34}$$

It is clear from Eqs. 34 and 20 that the expression of sensor output for both the modal wavefront sensor and the astigmatic defocus sensor are the same but with a variation in the expression of intensities. Now, let us find the expression of the sensitivity of the sensor.

4.1 Sensitivity of the Sensor

Since the expression of the sensor output of an astigmatic defocus sensor is identical to that of a modal wavefront sensor, the expression of sensitivity of the stigmatic defocus sensor will also be the same as that of the modal wavefront sensor. Thus, the expression of sensitivity of an astigmatic defocus sensor will be

$$\left.\frac{\partial O_A}{\partial a}\right|_{a=0} = \frac{2I_2 \frac{\partial}{\partial a} I_1\big|_{a=0} - 2I_1 \frac{\partial}{\partial a} I_2\big|_{a=0}}{(I_1 + I_2)^2\big|_{a=0}} \tag{35}$$

Similar to that of the modal wavefront sensor discussed in the previous section, we now have to find the expression of $\frac{\partial I}{\partial a}$ as

$$\begin{aligned}
\frac{\partial}{\partial a}\{I(u,v)\} &= \frac{1}{\lambda^2 f^2}\frac{\partial}{\partial a}\left|\int_0^{2\pi}\int_0^1 e^{jbZ_5(r,\theta)+jaZ_4(r,\theta)} e^{-j2\pi\rho r\cos(\phi-\theta)} r dr d\theta\right|^2 \\
&= \frac{1}{\lambda^2 f^2}\frac{\partial}{\partial a}\left[\int_0^{2\pi}\int_0^1 e^{j[bZ_5(r,\theta)+aZ_4(r,\theta)-2\pi\rho r\cos(\phi-\theta)]} r dr d\theta\right. \\
&\quad \left.\times \int_0^{2\pi}\int_0^1 e^{-j[bZ_5(r,\theta)+aZ_4(r,\theta)-2\pi\rho r\cos(\phi-\theta)]} r dr d\theta\right] \\
&= \frac{-j}{\lambda^2 f^2}\left[\int_0^{2\pi}\int_0^1 e^{j[bZ_5(r,\theta)+aZ_4(r,\theta)-2\pi\rho r\cos(\phi-\theta)]} r dr d\theta\right. \\
&\quad \times \int_0^{2\pi}\int_0^1 Z_4(r,\theta) e^{-j[bZ_5(r,\theta)+aZ_4(r,\theta)-2\pi\rho r\cos(\phi-\theta)]} r dr d\theta \\
&\quad - \int_0^{2\pi}\int_0^1 e^{-j[bZ_5(r,\theta)+aZ_4(r,\theta)-2\pi\rho r\cos(\phi-\theta)]} r dr d\theta
\end{aligned}$$

$$\times \int_0^{2\pi}\int_0^1 Z_4(r,\theta)e^{j[bZ_5(r,\theta)+aZ_4(r,\theta)-2\pi\rho r\cos(\phi-\theta)]}rdrd\theta \tag{36}$$

which results in

$$\begin{aligned}\frac{\partial}{\partial a}\{I(u,v)\}\bigg|_{a=0} &= \frac{-j}{\lambda^2 f^2}\left[\int_0^{2\pi}\int_0^1 e^{j[bZ_5(r,\theta)-2\pi\rho r\cos(\phi-\theta)]}rdrd\theta\right.\\ &\quad\times\int_0^{2\pi}\int_0^1 Z_4(r,\theta)e^{-j[bZ_5(r,\theta)-2\pi\rho r\cos(\phi-\theta)]}rdrd\theta\\ &\quad-\int_0^{2\pi}\int_0^1 e^{-j[bZ_5(r,\theta)-2\pi\rho r\cos(\phi-\theta)]}rdrd\theta\\ &\quad\times\int_0^{2\pi}\int_0^1 Z_4(r,\theta)e^{j[bZ_5(r,\theta)-2\pi\rho r\cos(\phi-\theta)]}rdrd\theta\end{aligned} \tag{37}$$

In this case also, at $b = 0$, both the sub-apertures I_1 and I_2 will have the same intensities. Further, an increase in a results in an increase in the intensity of one of the sub-apertures and a decrease in the intensity of the other. Thus, $\frac{\partial}{\partial a}(I_1) = -\frac{\partial}{\partial a}(I_2)$, therefore we have

$$\begin{aligned}\frac{\partial O_A}{\partial a}\bigg|_{a=0} &= \frac{2I_2\frac{\partial}{\partial a}I_1\big|_{a=0}-2I_1\frac{\partial}{\partial a}I_2\big|_{a=0}}{(I_1+I_2)^2\big|_{a=0}}\\ &= \frac{2I_1\frac{\partial}{\partial a}I_1\big|_{a=0}+2I_1\frac{\partial}{\partial a}I_1\big|_{a=0}}{(2I_1)^2\big|_{a=0}}\\ &= \frac{\frac{\partial}{\partial a}I_1\big|_{a=0}}{I_1\big|_{a=0}}\end{aligned} \tag{38}$$

Thus, Eq. 38 determines the sensitivity of the astigmatic defocus sensor. By applying Eqs. 32, 37 and 38, the sensitivity of the sensor can be calculated for various RMS amplitudes (b) of the Z_5 mode and different detector radii in Fourier plane coordinates (ρ_p).

Figure 6(a) shows the plot of the sensor's sensitivity for different RMS amplitude of Z_5 mode (b) and detector radius in Fourier plane coordinates (ρ_p). In the figure b is taken along the horizontal axis and ρ_p is taken along the vertical axis, while the sensitivity is represented in different gray scale ranging from $-$ 1.5 to $+$ 1.5. The physical detector radius can be calculated from the Fourier plane coordinates (ρ_p) by using the expression $R = \rho_p\lambda f$, where λ is the wavelength of the light beam and f is the focal length of the focusing lens used. It is observed that for RMS amplitudes of Z_5 ranging from $b = 0.8 - 1.4$ radian and $\rho_p > 0.8$ we get a comparatively good sensitivity (i.e. > 1). It is found that a sensitivity of ≈ 1.2 is achieved at $\rho_p = 1.4$. Figure 3(b) shows a plot of the sensitivity of the sensor versus RMS amplitudes of

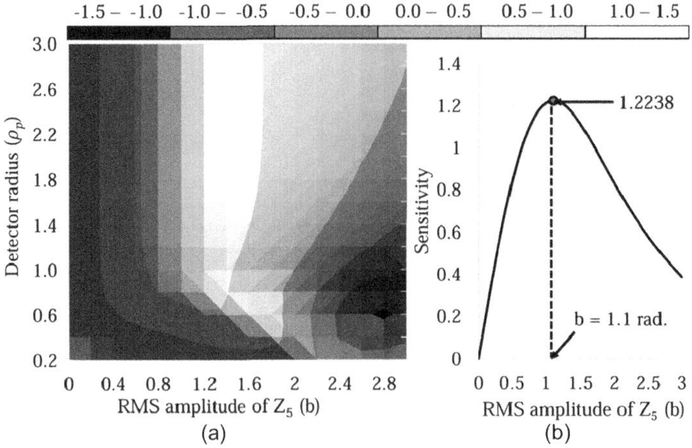

Fig. 6 Plot of sensitivity versus (a) RMS amplitude of astigmatism 45° (i.e. b) in radian and detector radius in Fourier coordinate (i.e. ρ_p) and (b) RMS amplitude of astigmatism 45° (i.e. b) in radian

Z_5 (b) taking $\rho_p = 1.4$. It is observed that the maximum sensitivity of 1.2065 is obtained at $b = 1.1$ radian.

4.2 *Cross-Sensitivity of the Sensor*

Similar to the modal wavefront sensor, the performance of the sensor is evaluated by assessing its cross-sensitivity in the presence of other Zernike modes (aberration modes) besides defocus. This involves examining the cross-sensitivity when one of the six low-order aberration modes is present alongside defocus in the beam's phase profile. Figure 7 illustrates the intensity distribution at the detector plane for a beam that includes astigmatism 0°(Z_6), *y* coma (Z_7), *x* coma (Z_8), trefoil *y* (Z_9), trefoil *x* (Z_{10}) and spherical aberration (Z_{11}), all when focused by an astigmatic lens.

From Fig. 7 it is clear that in the presence of astigmatism 0°(Z_6), *y* coma (Z_7), *x* coma (Z_8), trefoil *y* (Z_9) and trefoil *x* (Z_{10}), the sum of the total intensities of sub-aperture 1 & 3 (i.e. $I_1 + I_3$) is equal to the sum of the total intensities of sub-aperture 2 & 4 (i.e. $I_2 + I_4$). Thus, considering the expression of sensor output, i.e. $O_A = \frac{(I_1+I_3)-(I_2+I_4)}{(I_1+I_3)+(I_2+I_4)}$, the sensor output will be zero, i.e. $O_A = 0$. Thus, there will be no cross-sensitivity in the presence of these modes.

In the presence of spherical aberration (Zernike mode Z_{11}), the sum of intensities of sub-aperture 1 & 3 and the sum of intensities of sub-aperture 2 & 4 are not equal. However, it remains true that $I_1 = I_3$ and $I_2 = I_4$. Consequently, this situation is analogous to that of defocus (Zernike mode Z_4). Thus, there will be a non-zero cross-sensitivity, which can be determined by substituting the Zernike mode Z_4 by Z_{11} in Eqs. 32 and 37 and by using Eq. 38.

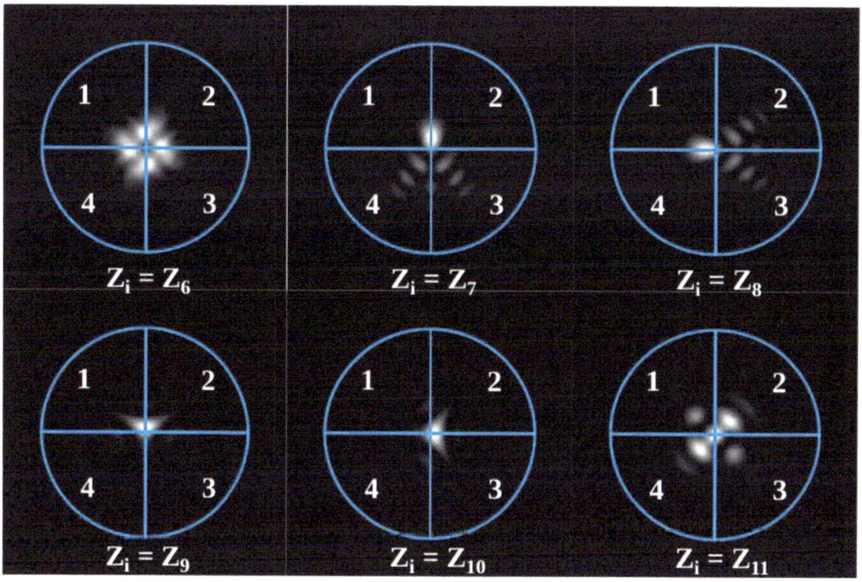

Fig. 7 Illustration of the intensity distribution at the detector plane for a beam that includes 1 rad RMS amplitude of astigmatism 0° (Z_6), *y* coma (Z_7), *x* coma (Z_8), trefoil *y* (Z_9), trefoil *x* (Z_{10}) and spherical aberration (Z_{11}), all when focused by an astigmatic lens

Table 3 Cross-sensitivity in an astigmatic defocus sensor caused by various Zernike modes

Zernike modes	Z_6	Z_7	Z_8	Z_9	Z_{10}	Z_{11}
Cross-sensitivity	0	0	0	0	0	– 0.497

Table 3 displays the cross-sensitivities associated with the Zernike modes Z_6, Z_7, Z_8, Z_9, Z_{10} and Z_{11}.

Table 3 shows that, aside from spherical aberration (Z_{11}), which exhibits a non-zero cross-sensitivity, all other modes (Z_6, Z_7, $Z8$, Z_9 and Z_{10}) shows zero cross-sensitivity. The cross-sensitivity for spherical aberration is measured as – 0.497, which is approximately 58% lower than the sensitivity for defocus aberration.

5 Comparison Between Both the Sensors

A comparison between the modal wavefront sensor and the astigmatic defocus sensor is tabulated below (Table 4).

Table 4 Comparison between modal wavefront sensor and astigmatic defocus sensor

Categories	Modal wavefront sensor	Astigmatic defocus sensor
Modes sensed	All the modes	Defocus only
Sensing ability	Low strength with high sensitivity; mid-strength with low sensitivity	Low strength with high sensitivity; high strength with low sensitivity
Bias aberration	Test mode itself	Astigmatism
Detector	Two-point size detectors are required	A quad detector or a multi-pixeled detector is required
Sensitivity	Maximum 0.8942 at 1.3 rad of Z_4	Maximum 1.2065 at 1.1 rad of Z_5
Cross-sensitivity	$-$ 0.227 in case of Z_{11}	$-$ 0.497 in case of Z_{11}

6 Conclusion

A detailed theoretical analysis of two bias-based techniques of measuring defocus has been carried out. The chapter started with a brief theoretical background on the calculation of the intensities of the back focal plane of a lens and an introduction to the Zernike polynomials and the displacement theorem. We then put forward the principles and theory of the two techniques, i.e. the modal wavefront sensor and the astigmatic defocus sensor. We also calculated the sensitivities and the cross-sensitivities of both the sensors. Lastly, we ended the chapter with a comparison between the two techniques.

References

1. Yan, R., Li, K., Jiang, W.: Defocus and magnification dependent variation of tem image astigmatism. Sci. Rep. **8**(1), 344 (2018)
2. Tokovinin, A., Kornilov, V.: Accurate seeing measurements with mass and dim. Mon. Not. R. Astron. Soc. **381**(3), 1179–1189 (2007)
3. Dinh, V.H., Vu, Y.N.T., Cao, X.B., et al.: Auto-focus methods in laser systems for use in high precision materials processing: a review. Opt. Lasers Eng. **167**, 107625 (2023)
4. Liu, J., Lu, Y., Huang, D., Yang, J., Fan, C., Chen, C., Li, J., Wang, Q., Li, S., Jiang, B., et al.: The efficacy of defocus incorporated multiple segments lenses in slowing myopia progression: results from diverse clinical circumstances. Ophthalmology **130**(5), 542–550 (2023)
5. Kumar, N., et al.: Experimental demonstration of in situ surface and thickness profile measurements of thin film during deposition using a grating array based wavefront sensor. Opt. Lett. **47**(21), 5509–5512 (2022)
6. Kohnen, T., Lemp-Hull, J., Suryakumar, R.: Defocus curves: focusing on factors influencing assessment. J Cataract Refract Surg **48**(8), 961–968 (2022)
7. Malik, K.S., Boruah, B.R.: Optimal trap velocity in a dynamic holographic optical trap using a nematic liquid crystal spatial light modulator. J. Opt. **24**(3), 034004 (2022)
8. Malik, K.S., Goutam Buddha, S.S., Kumar, N.: Calculation of optical forces in optical tweezers. Materials for electronic, magnetic, and spintronic technologies: characterization and applications from energy storage to disease detection. Cham: Springer Nature Switzerland, pp. 223–235. (2024)

9. Xiong, J., Hsiang, E.L., He, Z., Zhan, T., Wu, S.T.: Augmented reality and virtual reality displays emerging technologies and future perspectives. Light Sci. Appl. **10**(1), 1–30 (2021)
10. Zhang, Y., Shi, Z., Wang, H., Pang, J.: An efficient autofocus method for microscope based on the improved first-order derivative Gaussian filtering operator. Opt. Laser Technol. **170**, 110225 (2024)
11. Xu, R., Huang, Z., Gong, W., Zhou, W., Tropea, C.: Depth from defocus technique for irregular particle images. Measurement. 115156 (2024)
12. Rodríguez, G., Villa, J., Ivanov, R., González, E., Martínez, G.: Foucault test: a quantitative evaluation method. JOSA A **33**(8), 1604–1611 (2016)
13. Liu, Y., Li, J., Hu, Y., Zhang, S., Wang, Z., Xu, C., Yang, S., Zhao, W., Hao, Q.: Interferometric measurement of the radius of curvature based on axial displacement from a confocal position and corresponding defocus wave front. Appl. Opt. **62**(20), 5443–5451 (2023)
14. Guo, J., Han, S., Zhang, L., Shen, Y., Lv, W., Yang, Y., Gong, C., Han, M.: Analysis of defocus error on the absolute measurement of spherical surface. In: Optical Metrology and Inspection for Industrial Applications X. vol. 12769, pp. 211–218. SPIE (2023)
15. Cheng, H., Zhu, X., Liu, Y., Wang, L., Wu, Y., Shen, C.: Phase retrieval at all defocus distances. J. Opt. **51**(1), 184–193 (2022)
16. Kumar, N., Khare, A., Boruah, B.R.: Enhancement of wave front measurement sensitivity in a zonal wavefront sensor without curtailing the sensing speed. J. Opt. **24**(1), 014003 (2021)
17. Kumar, N., Khare, A., Boruah, B.R.: Enhanced dynamic range of the grating array based zonal wavefront sensor using a zone wise scanning method. Photon. Instrum. Eng. VII SPIE. 11287 (2020)
18. Kumar, N., et al.: Improvement in wave front measurement accuracy in grating array-based zonal wavefront sensors by modulating the intensity profile of the incident beam. Practical Holography XXXVII: Displays, Materials, and Applications, vol. 12445. SPIE (2023)
19. Platt, B.C., Shack, R.: History and principles of shack-hartmann wavefront sensing (2001)
20. Neil, M.A., Booth, M.J., Wilson, T.: New modal wave-front sensor: a theoretical analysis. JOSA A **17**(6), 1098–1107 (2000)
21. Konwar, S.: Astigmatic defocus sensor: its performance and optimization. J. Opt. 1–13 (2024)
22. Konwar, S., Sherpa, N.: Astigmatic lens approach of defocus measurement using computer generated hologram. Optik **309**, 171873 (2024)
23. Konwar, S., Kumar, N., Goutam Buddha, S.S.: Direct defocus measurement using single astigmatic lens: a theoretical study. Photon. Instrum. Eng. X. Vol. 12428. SPIE, (2023)
24. Aumann, S., Donner, S., Fischer, J., Müller, F.: Optical coherence tomography (oct): principle and technical realization. High-Resolut. Imag. Microsc. Ophthalmol. New Front. Biomed. Opt. 59–85 (2019)
25. Konwar, S., Boruah, B.R.: Estimation of inter-modal cross talk in a modal wavefront sensor. OSA Continuum **1**(1), 78–91 (2018)
26. Konwar, S., Boruah, B.R.: Improved linear response in a modal wavefront sensor. JOSA A **36**(5), 741–750 (2019)
27. Goodman, J.W.: Introduction to Fourier optics. 3rd, vol. 3. Roberts and Company Publishers. (2005)
28. Noll, R.J.: Zernike polynomials and atmospheric turbulence. JOsA **66**(3), 207–211 (1976)
29. Born, M., Wolf, E.: Principles of optics: electromagnetic theory of propagation, interference and diffraction of light. Elsevier (2013)

Advancements in Optical Fiber and Photonics Crystal Fibers

Ajay Kumar and Vikas Kumar

Abstract This chapter discusses the evolution from standard optical fibers to photonic crystal fibers (PCFs). Optical fibers are essential in modern telecommunications, transmitting light signals with minimal loss over long distances. PCFs, however, enhance performance by using a micro-structured arrangement of air holes within the fiber, allowing unique light guidance properties such as endlessly single-mode operation, high nonlinearity, and tailored dispersion control. The chapter covers the design, fabrication, and advantages of PCFs over conventional optical fibers, emphasizing their applications in telecommunications, high-power transmission, sensing, and nonlinear optics. It highlights the revolutionary potential of PCFs to provide unprecedented control over light propagation and innovative solutions across various fields. PCFs incorporate advanced optical materials called photonic crystals, which have periodic patterns that create photonic band gaps, preventing certain wavelengths from propagating. These structures give PCFs unique light-guiding qualities. Solid core PCFs offer high nonlinearity and dispersion control, while photonic band gap fibers allow low-loss transmission and precise light propagation control by confining light within a core using a photonic band gap. These advancements make PCFs invaluable for applications in nonlinear optics, sensing, and telecommunications. The chapter also covers computational techniques essential for designing and analyzing photonic crystals and PCFs, including the beam propagation method (BPM), finite element method (FEM), and full vector-plane wave expansion method (PWEM). BPM is used to model nonlinear effects and complex waveguide geometries, FEM provides detailed structural analysis by solving Maxwell's equations, and PWEM helps analyze light-guiding properties and predict photonic band gaps. These techniques enable comprehensive analysis, optimization, and innovation of advanced photonic devices, enhancing their performance in applications like quantum technologies, sensing, and telecommunications.

Keywords Photonic crystal · Photonic band gap fibers · Beam propagation method

A. Kumar (✉) · V. Kumar
Department of Physics, Indian Institute of Technology Delhi, New Delhi, India
e-mail: ajaykumargainpura@gmail.com

A. Mishra and V. Pathak (eds.), *Smart Materials for Energy Storage and Biomedical Applications*, Engineering Materials, https://doi.org/10.1007/978-3-031-92584-9_6

1 Introduction

Light, with its high frequency of approximately 10^{14} Hz, presents an opportunity for significantly higher modulation rates compared to radio or microwave frequencies, allowing for the transmission of substantial information within a single communication channel. To enhance the capacity for information transmission, it's crucial to have a large frequency bandwidth for each bit of information. This can be achieved by utilizing optical frequencies, which are notably high, around 10^{15} Hz, rather than relying on the lower frequencies of radio waves or microwaves. Thus, optical frequencies are a superior choice for transmitting information.

Optical fibers and photonic crystal fibers possess distinct properties that make them highly effective in transmitting vast amounts of data, which has heightened interest in their scientific and technological applications. The use of optical fibers as the guiding medium for this data transmission is ideal due to their extremely low loss rate (0.2 dB/km), which minimizes the need for repeaters in communication systems. Modern fiber optics systems have made it possible to transmit several hundred megabits of information per second, equating to around five hundred thousand independent voice channels [1]. As a result, fiber optics has proven to be a reliable technology, with system costs decreasing significantly over time.

In this context, this chapter explores the various types of optical fibers based on the number of modes they support. It also delves into fiber modeling for different applications and provides an analytical solution for the modal fiber profile within such structures. The chapter concludes by discussing recent and highly practical applications of fibers in both communication and non-communication fields. A schematic of a multimode fiber (MMF) with a step-index profile is also presented [2]. However, the propagation of light through the atmosphere is constrained by line-of-sight communication, which is susceptible to weather conditions. In 1870, Tyndall demonstrated that light could be guided within a water jet via total internal reflection. Despite theoretical exploration, it wasn't until the mid-1960s that light-based communication using circular dielectric waveguides gained serious attention. A significant breakthrough occurred in 1954 with the introduction of clad dielectric waveguides, which addressed previous limitations.

When an electromagnetic wave encounters the boundary between two dielectric media with differing refractive indices, part of the wave is reflected, and the rest is transmitted. According to Maxwell's equations, the tangential components of the electric and magnetic fields, along with the normal components of electric displacement and magnetic flux density, must remain continuous across this boundary. The relationships derived from these principles, known as Fresnel's equations, describe the behavior of reflected and transmitted waves relative to the angle of incidence. These equations determine the angles of the reflected and transmitted waves. Depending on the orientation of the electric field, the vector, whether parallel or perpendicular to the plane of incidence—waves are categorized as either Transverse Magnetic (TM) or Transverse Electric (TE). For TM polarization, the magnetic field remains perpendicular to the plane of incidence, while for TE polarization, the electric field remains perpendicular. Fresnel's equations offer a mathematical framework to understand these polarization effects and the energy distribution between reflected and transmitted waves at the boundary [1–3].

In Sect. 2, we discuss the basics of optical waveguide and optical fibers and in Sects. 3 and 4, different types of losses and dispersion are discussed. In Sects. 5 and 6, optical fibers in communication and sensors are discussed. Further in Sect. 7, we discussed various limitations of conventional optical fibers are discussed and comparison of PCFs with conventional optical fibers to name the advantages of PCFs is carried out. In Sect. 3, various important properties of PCFs have been discussed, for example their structural properties, guiding mechanisms, dispersion tailoring, nonlinearity tailoring, propagation and bend loss characteristics. In Sects. 8 and 9, various applications which can be offered by customized PCFs have been discussed. Finally, we conclude the chapter.

2 Optical Waveguides and Fibers

The planar symmetric dielectric waveguide is a simple yet essential form of waveguide, useful for understanding waveguide behavior and applicable to circular waveguides with minor modifications. It consists of dielectric-core with a higher refractive index n_1 sandwiched between two cladding regions with a lower refractive index n_2 (see Fig. 1a). Light propagates through the core in a zigzag pattern due to total internal reflection at the core-cladding interfaces. For this to occur, the internal ray angle must be between 90° and a critical angle. It is possible for an endless number of rays to propagate along the guide at the same time, each slightly displaced. These rays represent lines normal to the plane wavefronts. The key requirement for light propagation is that all points along a wavefront must have same phases. If the phase at different points along a wavefront differs by anything other than a multiple of 2π, destructive interference will occur, preventing light propagation. The phase-difference depends on the length of the i.e. path length and the phase changes due to reflections at the boundary of core-cladding interfaces. These phase changes vary depending on whether the radiation is Transverse Electric (TE) or Transverse Magnetic (TM). The total phase change is critical in determining the propagation of light within the waveguide.

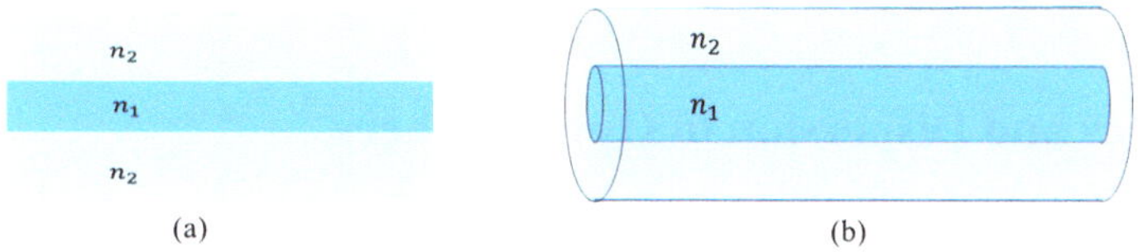

Fig. 1 Schematic of **a** planar optical waveguide and **b** optical fiber

Optical fibers typically have a cylindrical structure with a core surrounded by cladding. The core's refractive index (n_1) is more than the cladding index (n_2), leading to these being called step-index fibers due to the step-like refractive index profile (see Fig. 1b). An additional protective coating, usually plastic, is often added to strengthen the fiber and guard against chemical damage. Like planar waveguides, light within the fiber is confined to certain 'modes,' each representing a specific electromagnetic field pattern. However, the mathematical equations describing these modes in circular waveguides are more complex compared to those in planar waveguides. In circular waveguides, such as optical fibers, light travels in two types of rays: meridional and skew rays. Meridional rays pass through the central axis of the fiber, like rays in planar waveguides, and give rise to 'Transverse Electric (TE)' and 'Transverse Magnetic (TM)' modes. However, in circular waveguides, two integers, l and m, are required to fully describe these modes, leading to designations like TE_{lm} and TM_{lm}. This distinction arises because circular waveguides have a two-dimensional structure, unlike the one-dimensional planar waveguides. Skew rays, which have no counterpart in planar waveguides, travel in helical paths around the fiber axis. These rays generate modes classified as HE_{lm} or EH_{lm}, depending on whether the magnetic or electric field components are dominant. Some skew rays, known as leaky modes, do not correspond to bound modes but can still propagate over considerable distances before being lost [3, 4].

In most practical optical fibers, the refractive indices of the core and cladding differ by only a few percent. This small difference allows the complex set of modes (TE_{lm}, TM_{lm}, HE_{lm}, and EH_{lm}) to be approximated by a simpler set of linearly polarized (LP_{lm}) modes. These LP modes have specific electric field intensity profiles, characterized by field maxima along a radius vector and around the fiber's circumference. The l value indicates the degree of helical propagation, with larger l values corresponding to tighter helices. The l-value is related to the angle of the rays, with larger m values corresponding to smaller angles [3].

In multimode fibers, the number of modes is so large that they form a near-continuum of ray angles within the guide. This allows for an analytical approach to understanding some of the fiber's properties using ray theory. However, when the number of modes is very large or when only one mode can propagate, the analysis becomes more complex, requiring a more detailed understanding of the waveguide's behavior. Overall, the behavior of light in circular waveguides involves a more intricate interplay of modes compared to planar waveguides, necessitating a deeper analysis to fully comprehend the propagation characteristics within optical fibers.

3 Losses and Dispersion in Optical Fibers

When optical fiber is used for communication purposes, it is expected to transmit light signals between its two ends without any loss of strength [1, 3]. However, in practice, the fiber does cause signal attenuation due to various types of losses, which are illustrated in Fig. 2.

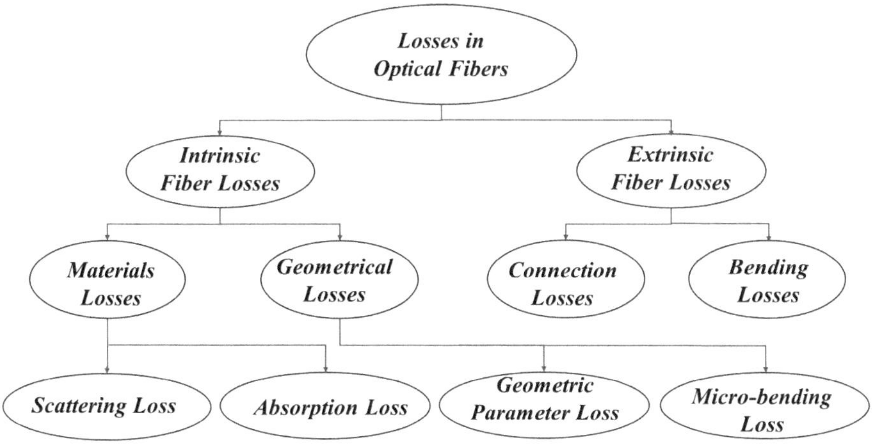

Fig. 2 Different types of losses in optical fibers

Key points to note include:

- Although 'optic' in the electromagnetic spectrum refers to frequencies in the infrared, visible, and ultraviolet bands, the visible and ultraviolet portions suffer from significant attenuation. As a result, effective optical communication occurs primarily in the infrared band, where losses are comparatively lower.
- Even within the infrared band, not all of it is suitable for communication. The presence of hydroxyl (OH^-) ions dissolved in the glass causes strong absorption losses in the fiber. These ions create sharp absorption peaks in the silica absorption spectrum, but there are narrow windows between these peaks that are unaffected by OH^- absorption. The three best windows for optical fiber communication are identified at 850 nm, 1310 nm, and 1550 nm, known as the first, second, and third telecommunication windows, respectively. Most optical fiber communication systems operate within one of these three windows.
- The refractive index inside the glass is not uniform due to imperfections introduced during manufacturing. These imperfections cause random molecular locations that scatter light passing through the fiber, a type of loss known as Rayleigh scattering.
- Signal attenuation in the fiber is a critical parameter because it determines the maximum distance over which optical communication can occur without the need for a repeater. The attenuation of a signal, measured in decibels (dB), is defined by the equation $A = 10 \log (P_{in}/P_{out})$, where P_{in} and P_{out} represent the input and output powers of the signal, respectively.

4 Dispersion

When optical fibers are used for communication, they are expected to transmit light signals without any loss of strength. However, in practice, signal attenuation occurs due to several factors. Effective communication primarily happens in the infrared spectrum, where losses are lower, specifically at 850, 1310, and 1550 nm, known as the first, second, and third telecommunication windows. These are less affected by hydroxyl ion absorption in fiber. Additionally, imperfections in the fiber's structure led to Rayleigh scattering, further contributing to signal loss. Dispersion, which includes pulse (or modal) and material (or chromatic) dispersion, also impacts the signal by causing pulse spreading, limiting the bandwidth and data transmission capacity. Both attenuation and dispersion are critical in determining the maximum transmission distance and information-carrying capacity of fiber without the need for repeaters. Light pulses disperse as they pass through the optical cable due to a process called dispersion. This spread can cause signals to overlap, which can deteriorate the sent data quality, especially in high-speed communication networks. Waveguide dispersion and chromatic dispersion are the two main types of dispersion seen in optical fibers. Because different light wavelengths go through fiber materials at various speeds, chromatic dispersion happens. Pulse broadening results from the various parts of a light pulse arriving at their destination at different times due to this speed difference. Conversely, waveguide dispersion results from the physical characteristics of the fiber, particularly the size of the cladding and core, which influence the propagation of various light modes [1–4].

Another kind of dispersion in optical fibers is called waveguide dispersion, and it is impacted by the geometry of the fiber, specifically the diameters of the cladding and core. The overall dispersion characteristics of the fiber can be influenced by optimizing the fiber's architecture to regulate the propagation of various light modes. One type of fiber that is intended to balance chromatic and waveguide dispersion is called non-zero dispersion-shifted fiber (NZDSF). Because these fibers are designed to have a tiny but non-zero dispersion at the operating wavelength, nonlinear phenomena like four-wave mixing (FWM) can be avoided, which can happen with zero dispersion fibers. NZDSF fibers can sustain high data rates while reducing the possibility of signal distortion and interference by keeping a small degree of dispersion.

Another important component affecting optical fiber performance is optical loss, often known as attenuation. The term 'loss' describes the weakening of the signal as it passes through the fiber. Several things, such as Rayleigh scattering, fiber bending, and absorption of contaminants in the fiber material, can contribute to this reduction. In silica-based fibers, especially at shorter wavelengths, Rayleigh scattering—during which the density of the fiber material varies microscopic—is the primary cause of loss. Light is scattered in various directions because of interacting with these variances, which weakens the signal.

Maintaining fiber-optic communication systems' efficiency requires controlling optical loss. To decrease impurities and structural flaws in optical fibers and lessen the effect of absorption and scattering losses, high-purity materials and sophisticated production procedures are employed. Furthermore, regulating loss over long distances requires the use of optical amplifiers. 'Erbium-Doped Fiber Amplifiers (EDFAs)' can increase signal intensity without the need for electronic regeneration, and they are especially effective in the 1550 nm wavelength region. By boosting the light signal's energy as it travels through the fiber, these amplifiers make up for transmission losses and enable longer signal transmissions without noticeably degrading the signal [2, 3].

The ability of optical fibers to efficiently control loss and dispersion makes them perfect for high-end communication systems. Low chromatic dispersion in these fibers lessens pulse broadening and permits higher data transmission speeds. Additionally, they have minimal attenuation, which enables messages to go farther without noticeably losing intensity. These characteristics are particularly crucial for systems that employ dense wavelength-division multiplexing (DWDM), a method that boosts a system's overall data-carrying capacity by enabling the simultaneous transmission of many wavelengths across a single fiber [3].

The continued development of new optical fiber technologies with improved dispersion and loss management capabilities is being driven by the need for quicker and more dependable communication. The creation of hollow-core fibers is among the most exciting developments in this industry. Hollow-core fibers direct light via an air-filled core as opposed to conventional fibers, which direct light through a solid glass core. Since light moves through air more efficiently than through glass, this design may result in much less attenuation and dispersion. Additionally, hollow-core fibers are less prone to nonlinear effects, which makes them a desirable choice for high-speed communication systems.

4.1 Pulse Dispersion

A popular method for controlling chromatic dispersion is the application of single-mode fibers. Because of the single mode of propagation allowed in these fibers, intermodal dispersion is greatly reduced. Single-mode fibers guarantee that light pulses stay more firmly restricted by restricting the propagation to a single mode, therefore decreasing the possibility of pulse broadening. Because of this, single-mode fibers are especially well suited for long-range communication, in which keeping signal quality requires limiting dispersion.

Dispersion-shifted fibers are another tactic for controlling chromatic dispersion (DSF). These fibers are designed with a zero-dispersion wavelength that is near the system's operating wavelength. Dispersion-shifted fibers lessen transmission-related pulse broadening by lining up the zero-dispersion wavelength with the working wavelength. Because of its great data-carrying capacity and low attenuation, the 1550 nm wavelength range is widely utilized in telecommunications. For these systems, this method is very helpful.

To mitigate the effects of chromatic dispersion in normal fibers, dispersion-compensating fibers (DCF) are frequently utilized in addition to dispersion-shifted fibers [2]. Positive dispersion accumulated in the transmission fiber can be reversed by dispersion-compensating fibers since they are built with negative dispersion. The entire dispersion can be balanced by sequencing a length of dispersion-compensating fiber after a segment of standard fiber. This effectively cancels out the pulse broadening and restores the signal quality. To sustain high data rates over long distances, long-haul fiber-optic communication systems frequently adopt this technology.

4.1.1 Pulse Dispersion in Step Index Fibers

In a multimode step-index fiber, different modes of light, though entering simultaneously, travel different path lengths within the fiber. As a result, pulses launched in these modes arrive at the fiber's end at different times, producing pulse stretching over time, known as modal or multi-path dispersion. This effect is particularly significant in step-index multimode fibers. Pulse broadening limits the fiber's transmission capacity by causing overlapping of consecutive pulses, making them unresolvable. To prevent overlap, the time delay between pulses must be increased, which reduces the number of pulses transmitted per unit time, thereby lowering the fiber's transmission capacity [1].

4.1.2 Pulse Dispersion in Graded Index Fibers

The graded index (GRIN) fiber has a core where the refractive index gradually decreases as the light ray moves away from the fiber's axis. This parabolic refractive index profile helps reduce pulse dispersion. In GRIN fibers, modes further from the core center travel faster, causing light pulses in different modes to reach the fiber's end at nearly the same time, minimizing dispersion. As a result, GRIN fibers experience less pulse broadening, allowing more data to be communicated with minimal distortion. The information-carrying capacity is mainly limited by material dispersion [2].

4.2 Material Dispersion

We've discussed pulse broadening in optical fibers due to light rays entering at different angles, causing them to take varying amounts of time to travel the same distance. GRIN fibers help reduce this type of dispersion. However, even if pulse broadening from this cause is eliminated, it doesn't mean the beam is entirely free from it. Another source of pulse broadening is material dispersion, which occurs because the refractive index varies with the light's wavelength. Since light from a source consists of multiple wavelengths, each travel at a different speed, causing them to arrive at different times, leading to pulse broadening. This type of dispersion, also known as chromatic dispersion, is more significant in single-mode fibers [3].

5 Optical Fiber Communication

The foundation of contemporary communication networks is made up of optical fibers, which allow enormous volumes of data to be transmitted over great distances quickly and with little loss. The fibers consist of a core that guides light and a cladding that surrounds it. The cladding has less refractive index than the core, which ensures that light stays contained within the core due to total internal refraction or (TIR). Dispersion and loss are two important issues that greatly affect the performance of optical fibers, even though they have revolutionized communications. Sustaining the accuracy and effectiveness of data transfer requires careful management of these variables.

Particularly in long-distance communication systems where data must travel over hundreds or even thousands of kilometers, chromatic dispersion is a major concern in optical fibers. Light pulses can overlap with one another as they propagate, making it challenging to discern between distinct signals. This may lead to data transmission problems, which would lower the communication system's overall performance. Optical fibers are engineered with specific qualities that aid in limiting pulse broadening to mitigate the effects of chromatic dispersion [5]. Enhancing the performance of fiber-optic communication networks is anticipated to be largely dependent on developments in machine learning and signal processing methods in addition to new fiber architectures. These technologies can further improve the communication system's efficiency and reliability by assisting in the detection and correction of faults brought on by dispersion and loss through real-time analysis of the data transmitted through fiber. Additionally, dispersion and loss effects can be anticipated and mitigated in advance by machine learning algorithms, enabling more proactive control of the fiber network. Theoretically, optical communication systems' capacity can be enhanced using OAM-MDM, but crosstalk with mode coupling in FMF or MMF limits OAM transmission; a special ring-core fiber suppresses this [5, 6].

5.1 Orbital Angular Momentum Modes: Ring Core Fibers

OAM modes provide a new basis for representing spatial modes, attracting growing interest in specialty fibers. This paper presents a comprehensive characterization of OAM modes in conventional graded-index multimode fiber (MMF). In this study,: (1) two degenerate cylindrical vector modes (eigenmodes) are combined to create circularly polarized OAM modes, and total angular momentum, including spin and orbital angular momentum, is analyzed; (2) all OAM modes in conventional graded-index MMF are divided into 10 mode groups, with effective refractive index differences between groups above 10^{-4}, allowing for low inter-group crosstalk; (3) chromatic dispersion, differential group delay, effective mode area, and nonlinearity are examined for each OAM mode group over a wavelength range of 1520–1630 nm, encompassing the entire C band and L band; (4) a test of performance tolerance to bending and fiber ellipticity; and (5) a test of performance resilience to fiber disturbances, such as core size, index contrast, and the imperfect. These findings provide a theoretical foundation for further space-division multiplexing applications using OAM modes in conventional graded-index fibers [7, 8].

A new degree of freedom for light waves is provided by vortex beams carrying orbital angular momentum (OAM), which enhances more conventional characteristics including intensity, phase, frequency, duration, and polarization. The problem of limited information capacity can be resolved by OAM's potentially infinite orthogonal states. OAM router devices, or OAM-RDs, are now essential for increasing the adaptability and usefulness of communication networks due to developments in OAM optical communication technology. This analysis assesses the state of technology today and highlights significant advancements in OAM-RDs. Additionally, it divides current studies on OAM-RD approaches into five main categories: adding/extracting, filtering, hopping, channel switching, and channel multicasting. The concepts, research methodologies, advantages, and disadvantages are also covered in detail, and future development trends and possibilities for OAM-RDs are examined [7, 9].

6 Fiber Optics Based Sensors

In industrial processes, monitoring parameters such as displacement, pressure, temperature, flow rate, liquid level, and chemical composition is crucial. The optimal measurement method should be sturdy, dependable, impervious to corrosion, inherently safe, and unaffected by outside influences. These measurements could benefit greatly from the use of optical techniques and fibers. However, established alternative technologies often dominate, and it may take time for optical fibers to make a significant market impact.

Optical fiber sensors are categorized into two main types: intrinsic and extrinsic. In extrinsic sensors, the optical fiber serves as a transmission channel, with modulation occurring outside the fiber. In intrinsic sensors, the measured directly affects the fiber, modifying the radiation passing through it. The radiation's amplitude, phase, or polarization can all be altered by intrinsic sensors. Amplitude modulation is the most practical choice in multimode fibers because phase or polarization modulation cannot be used due to mode coupling and erratic relationships between phases and polarization states. Conversely, single-mode fibers enable both phase and polarization modulation, allowing for the development of fiber-based interferometric sensors with exceptionally high sensitivities. Therefore, it is essential to consider multimode and single-mode intrinsic fiber sensors separately.

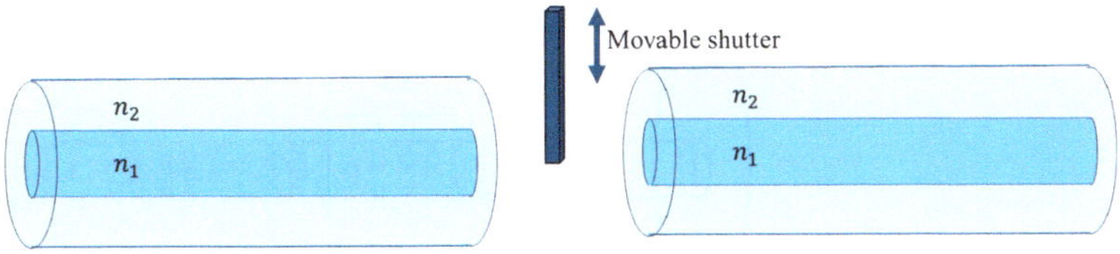

Fig. 3 Schematic of motion/displacement sensors using a shutter

6.1 Motion/Displacement-Sensors

Some of the very simple extrinsic fiber-optics based sensors are designed to measure movement or position. As an example, when two fiber ends are misaligned, the coupling loss varies depending on the displacement. The relationship between loss and displacement has been previously analyzed. A similar sensor employs a shutter moving between two laterally displaced fiber ends, where sensitivity can be improved by adding a pair of gratings, one fixed and the other movable [10]. However, while sensitivity increases, the measurement range decreases due to the periodic output related to the grating spacing. The movement range in a single-shutter sensor is constrained by the fiber core diameter, but using a beam expander between the fibers can significantly extend this range (Fig. 3).

One of the first generally available displacement sensors, the 'Fotonic sensor,' utilizes a fiber bundle, with half connected to a radiation source and the other half to a detector. When placed near a reflective surface, light reflects into the detecting fibers, with the detected amount depending on the distance from the fibers to the surface. Analyzing this setup with two fibers reveals that at very close distances, no light is coupled, but as the distance increases, so does the coupled light, until the detecting fiber is fully illuminated. Beyond this point, output decreases with distance, eventually following an inverse square law.

6.2 Few-Mode Fiber Bragg Grating

The most important structure in this category is the fiber Bragg grating (FBG). By embedding the FBG in a single-mode fiber, FBGs have been used as single-parameter sensors to monitor strain and temperature separately in recent years. Two separate measurements are needed to measure temperature and strain at the same time, and this method can be expanded to measure more than two parameters. Few-mode FBGs allow for such measurements. By etching the cladding down to the core of an FBG, an FM-FBG is created, which can simultaneously measure temperature, strain, and refractive index (Fig. 4).

FM-FBGs are generally azimuthally symmetric on the basis of refractive index that varies periodically along the propagation direction. The refractive index profile of such an FBG can be described as follows:

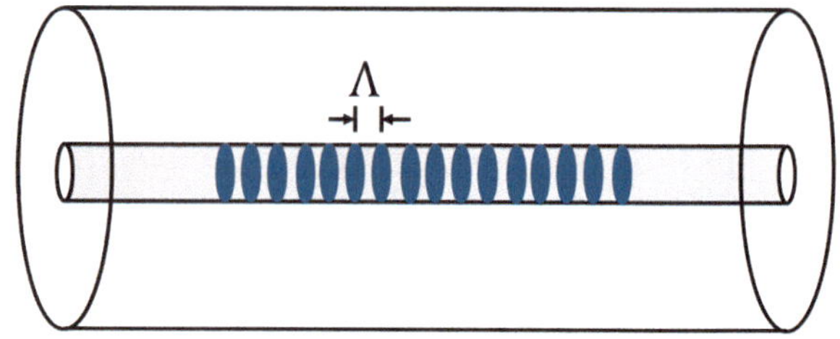

Fig. 4 Schematic of few-mode fiber Bragg grating

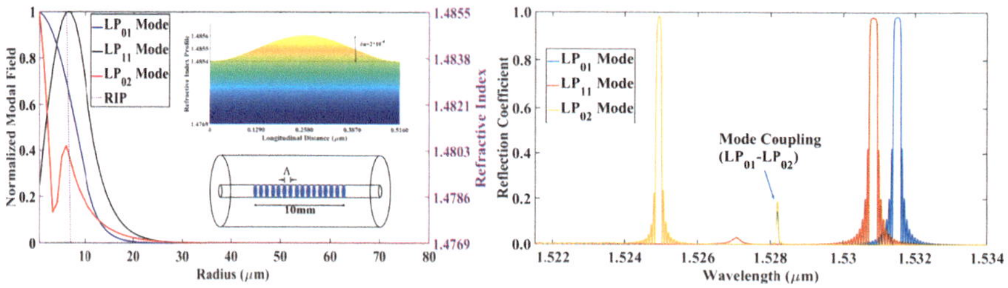

Fig. 5 **a** Modal field of different LP_{lm} modes, **b** reflection spectra of different modes in FM-FBGs

$$n(r,z) = \begin{cases} n_1 + \frac{\delta n}{2}\left(1 - \cos\left(\frac{2\pi z}{\Lambda}\right)\right) & r < r_c \\ n_2 & otherwise \end{cases}$$

In this context, $n_1 = 1.4855$ and $n_2 = 1.4769$ are the core and cladding refractive indices, with $\delta n = 2 \times 10^{-4}$ representing the index modulation in one period in the core. The core radius (r_c) is 6.3 μm, the grating period 0.516 μm at a wavelength 1.55 μm, and the total grating length is 10 mm. This FM-FBG supports only the first three modes: LP_{01}, LP_{11} *and* LP_{02}. Numerical modeling is conducted using 200 collocation functions, and propagation is simulated using the SSNP method to obtain the 'reflection' and 'transmission' coefficients for each wavelength [11–13].

The step index FM-FBG refractive index profile and the modal field intensity change with radial distance for various LP_{lm} modes are shown in Fig. 5a. A schematic of the FM-FBG and the refractive index cosine modulation for two grating periods are shown in the insets. For each mode's excitation, equal power is assumed, resulting in equal reflection peak power (height) for each mode as shown in Fig. 5b. A small peak due to mode coupling between the LP_{01} and LP_{02} modes are also observed between the reflection peaks corresponding to these modes, as expected.

6.3 Long Period Grating Sensors

Fiber optic technology has emerged as a promising alternative for detecting even minute amounts of chemical, gaseous, and biological substances. Fiber optic sensors

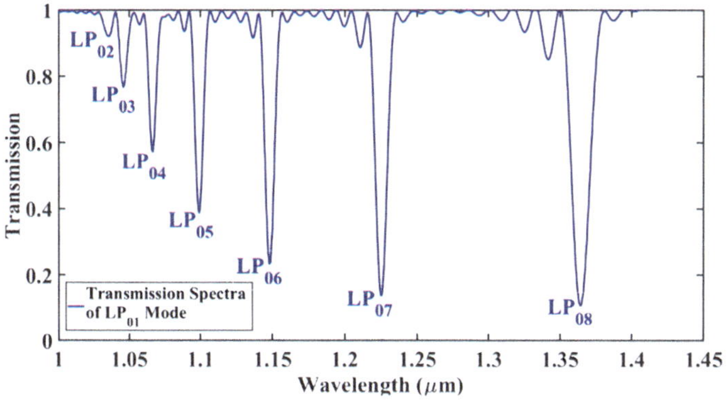

Fig. 6 Transmission spectra of LPG with collocation method using 100 collocation points

(OFSs) offer several advantages over traditional gas sensors, including such as portability, resistance to electromagnetic interference (EMI), the ability to be easily multiplexed, and the capability for long-distance monitoring. Among the various types of OFSs, Long Period Gratings (LPGs) have gained significant traction in various sensing applications, particularly for physical, chemical, and biological sensing. Since their introduction in the late 1990s, LPGs have become popular due to their high sensitivity to changes in temperature, strain, and surrounding refractive index (SRI) [14]. These sensors consist of a periodic perturbation in the refractive index or geometry of a section of optical fiber, typically measuring between 10 and 50 mm in length, with periods ranging from 100 to 1000 μm. LPGs can be fabricated using various techniques, including ultraviolet (UV) irradiation, CO_2 lasers, infrared (IR) femtosecond lasers, and electric arc discharge (Fig. 6).

Overall, these advancements highlight the potential of fiber optic technology, particularly LPG-based sensors, in enhancing the safety and effectiveness of monitoring systems for hazardous substances during transportation and in various industrial applications. The refractive index is measured by shifting the dip corresponding to highest LP_{0m} order mode.

7 Photonic Crystal Fibers

With their distinctive structural design and exceptional optical qualities, Photonic Crystal Fibers (PCFs) are a revolutionary advancement in the optical fiber industry [15–18]. In contrast to traditional optical fibers, which use a solid core encircled by cladding to direct light using total internal reflection, PCFs feature a recurring pattern of tiny air holes extending the length of them. With its remarkable control over light propagation, this photonic crystal structure opens new avenues for a wide range of photonics applications. Based on their guiding principles, PCFs are primarily classified into two groups: index guiding and photonic bandgap guiding.

The necessity to get past the drawbacks of conventional optical fibers led to the development of PCFs. It is difficult for traditional fibers to produce single-mode functioning over a wide wavelength range, but PCFs can do so by carefully controlling the size and configuration of the air holes [18]. This capacity is especially beneficial in the telecommunications industry, as reliable and low-loss signal transmission is essential for effective data transfer. Furthermore, the special structure of PCFs makes it possible to largely modify the dispersion characteristics, resulting in the production of fibers with flattened or zero dispersion over a broad range of spectral bands [19, 20]. This quality is crucial for controlling the propagation of pulses in ultrafast optics, where it's important to minimize distortion and preserve pulse shape.

Additionally, PCFs show remarkable nonlinearity, which makes them the perfect choice for producing supercontinuum light sources [21, 22]. These sources generate a wide spectrum of light from a narrowband input, which has important ramifications for spectroscopic techniques and medical imaging applications like optical coherence tomography (OCT). Furthermore, PCFs are particularly sensitive to changes in the environment because they can allow gases or liquids to seep into their air holes. This opens the door for enhanced sensing applications [23, 24]. Chemical and biological sensing take advantage of this sensitivity to provide strong instruments for industrial process control, medical diagnostics, and environmental monitoring.

PCFs are highly advantageous in the field of fiber lasers and high-power laser delivery [25]. They are appropriate for industrial uses such laser machining, cutting, and welding because of their huge mode area, which can withstand greater power levels without suffering damage. Furthermore, by enabling high power levels and effective light amplification, PCFs improve the performance of fiber amplifiers used in scientific and medical applications. All things considered; photonic crystal fibers have completely changed the field of optical fiber technology. Advancements in sensing, medical imaging, high-power laser applications, telecommunications, and other fields have been made possible by their diverse optical qualities and distinctive design. It is anticipated that as PCF research and development continue, its influence on industry and technology will grow, leading to more advancements in the field of photonics.

In summary, photonic crystal fibers (PCFs) are categorized based on their guiding mechanisms into index-guiding and photonic bandgap-guiding types. Index-guiding PCFs rely on total internal reflection with a micro structured cladding that enhances light confinement, while photonic bandgap fibers use a bandgap structure to confine light, even in a hollow core. Highly nonlinear PCFs, typically based on index-guiding principles, are designed with small cores to increase optical intensity and enhance nonlinearity. These fibers offer significant flexibility in dispersion, nonlinearity, and attenuation, making them valuable in various optical applications. An example of a commercially available highly nonlinear PCF demonstrates dispersion-flattened characteristics and a high nonlinear coefficient, showcasing the practical benefits of this technology in optical systems.

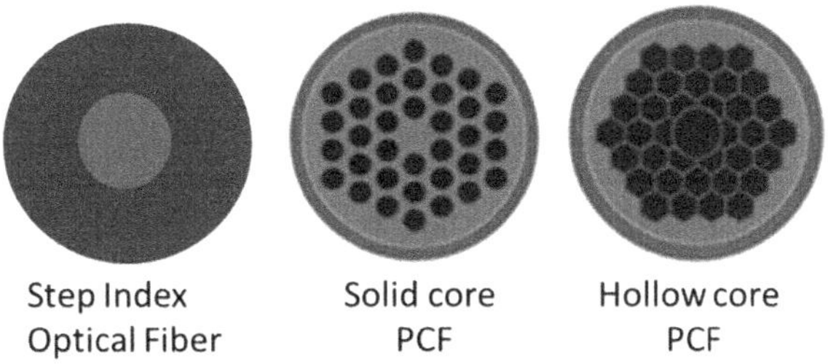

Fig. 7 Schematic of cross section of various optical fibers, the same is preserved along the length of fiber

Limitations of Conventional Optical Fibers and Advantages of PCF

Despite being extensively utilized in sensing and telecommunications applications, traditional optical fibers have several drawbacks. They have nonlinear effects and dispersion, which can deteriorate signal quality at high power levels and across long distances [26]. Standard fibers' core and cladding materials also restrict the wavelength range that can be used for effective transmission, usually to the near-infrared spectrum [27]. Furthermore, they may exhibit signal changes due to their sensitivity to external conditions including strain, temperature, and pressure [28]. Their limited adaptability stems from the fact that the waveguide properties cannot be customized for applications [29]. Additionally, standard fibers have a relatively significant bend loss, which makes them less appropriate for intricate routing and tight bending applications [30]. Figure 7 shows the cross sections.

PCFs, which incorporate a periodic microstructure, typically in the form of air holes, within the fiber cladding, alleviate many of the drawbacks of conventional optical fibers. This topology gives the fiber remarkable optical property control [15–20]. PCFs can be made to operate in a single mode across a broad range of wavelengths, including ultraviolet and visible light. They can be designed for dispersion management and have low dispersion, both of which are essential for high-speed data transfer [18–21]. PCFs can accommodate high-power applications and have low nonlinearity [21]. They can also continue to function in challenging conditions and are less susceptible to outside disturbances. A major benefit over conventional optical fibers is the novel uses in sensing, nonlinear optics, and high-power delivery made possible by PCFs' distinct structure [16, 22, 23].

8 Properties of PCFs

Fiber optic PCFs are a new advancement in the area that stand out for their superior performance attributes and distinct structural features. The primary characteristic of PCFs is their microstructured cladding, which is typically made up of a regular pattern of air holes spaced throughout the fiber's length. PCFs have several advantages over traditional optical fibers thanks to this structure, which allows for the fine tuning of a variety of properties for applications.

8.1 Structural Properties

PCFs are distinguished by their cladding, which is usually made up of a solid or hollow core surrounded by a hexagon-shaped pattern of air holes. Customizing the optical properties of the fiber is possible through exact control of the size, shape, and placement of these air holes during manufacture [17]. The core might be an expanded air hole (hollow-core PCF) or a solid silica strand (solid-core PCF), each of which has unique benefits.

8.2 Guiding Mechanisms

There are two main ways that PCFs can guide light: bandgap guiding and index-guiding. Similar to conventional fibers, index-guiding PCFs have a higher refractive index core but offer better control because of the surrounding air holes [18, 19]. Photonic bandgap fibers, or bandgap-guiding PCFs, use the periodic structure to provide a photonic bandgap that confines light to a hollow or lower refractive index core [30]. This makes it feasible to guide light in wavelength ranges that traditional fibers cannot or will not allow.

8.3 Nonlinear Optical Properties

The great intensity of light that may be contained in the tiny core of PCFs results in enhanced nonlinear optical characteristics [31, 32]. For nonlinear optics applications like four-wave mixing [33], soliton propagation [20, 21], and supercontinuum creation [20, 21], this makes them perfect. The air-silica microstructure's tight confinement and high effective nonlinearity are responsible for PCFs' high nonlinear coefficient. One of the examples of tiny core PCF for acquiring nonlinear characteristics is shown in Fig. 8. Given the core material is nonlinear.

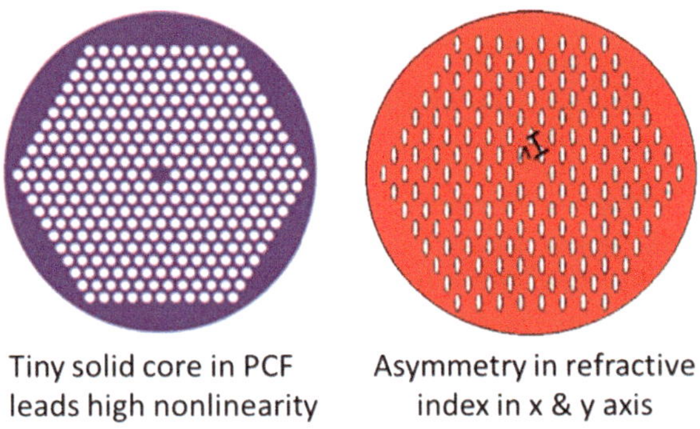

Fig. 8 Varieties of PCF leads to various applications, asymmetry of refractive index in x and y direction leads to birefringence

8.4 Birefringence and Polarization Control

High birefringence can be engineered into PCFs, which is advantageous for applications that need to maintain polarization. Strong birefringent qualities can result from the asymmetry created by changing the size or form of the air hole (as shown in Fig. 8, right PCF), which enables the fiber to sustain the polarization state of light over extended distances [34]. In this figure the air holes are elliptical in shape which leads to birefringence. This is especially helpful for polarized light distribution in telecommunications and sensing applications.

8.5 Attenuation and Losses

PCFs can be engineered to decrease bend loss in addition to exhibiting low attenuation comparable to conventional fibers [35]. The microstructured coating gives designers more freedom to bend fibers sharply without losing a lot of signals, which makes PCFs ideal for intricate and compact routing in photonic devices.

Beyond the capabilities of conventional optical fibers, PCFs provide a flexible platform with tunable optical properties. They are an effective tool in both research and industry because of their capacity to control dispersion, maximize nonlinear effects, maintain polarization, and minimize losses. They are also well-suited for advanced sensing applications. The spectrum of applications for PCFs is projected to increase as fabrication techniques advance, spurring additional advancements in fiber optics and photonics.

9 Applications of PCFs

9.1 Dispersion Tailoring in Solid Core Photonic Crystal Fibers (PCFs) and Applications

For many optical applications, solid core PCFs provide outstanding control over dispersion parameters. Customized dispersion in solid core PCFs is made possible by the microstructured cladding, which is made up of air holes positioned sporadically around a solid silica core. It is possible to accurately alter the cladding's effective refractive index by adjusting the size, shape, and spacing of these air holes. This makes it possible to tailor the dispersion profile of the fiber to produce desired effects, such anomalous dispersion, flattened dispersion for wide wavelength range, or zero dispersion at wavelengths [18, 19]. Applications like supercontinuum generation, where a broad and flat spectrum is required, and high-speed optical communications, where minimizing pulse broadening is crucial, benefit greatly from these tuned dispersion features [20, 21]. Furthermore, in ultrafast optics, dispersion control in PCFs is essential for producing and distributing femtosecond pulses with the least amount of distortion [36].

A computational method called the Full Vector Beam Propagation Method (FV-BPM) is employed to accurately predict light propagation in solid core PCFs [37]. FV-BPM provides a thorough study of intricate waveguide structures such as PCFs since it considers all vector components of the electric and magnetic fields, in contrast to scalar or reduced vector approaches. The nuances of light interaction with the micro structured cladding of solid core PCFs, such as polarization effects, modal dispersion, and nonlinear interactions, are especially well captured by this technique. FV-BPM enables accurate modeling of how changes to the fiber structure will affect its optical properties, which is crucial for designing and improving PCFs for applications. Researchers and engineers may create more effective and customized fiber designs by using FV-BPM to forecast how PCFs will function in practical applications including high-power lasers, nonlinear optical devices, and sensing systems.

We have designed a specialized solid core polyethylene PCF for THz wave propagation, the schematic of that is shown in Fig. 9a. Here, the pitch, Λ is called distance between the centers of air holes which is considered as 300 μm. As the wavelength range of 1 THz electromagnetic wave frequency is 300 μm, one can easily understand that the structural parameters, especially pitch are of the same range so as to get tailoring of dispersion versus wavelength curve. The large air holes' diameter, d_1 is 0.8 times pitch, they have been considered so that the proper confinement of the mode inside the core can be achieved (see Fig. 9b). The small area holes' diameter, d_2 is 0.3 times pitch, they have been considered so that large mode area can be achieved inside the PCF which can be used for high power delivery. The effective index versus wavelength data has been calculated using full vector beam propagation method (Beamprop module in RSoft, optical software by Synopsis). The dispersion versus wavelength curve (shown in Fig. 9c) is calculated from the effective index versus wavelength curve using the formula shown below.

$$D = -\frac{\lambda}{c}\frac{d^2\eta_{eff}}{d\lambda^2}$$

These air holes selection is optimized such that the dispersion versus wavelength curve can have two zero dispersion wavelengths as shown in Fig. 9c. Also, the dispersion and propagation less (as the polyethylene has very high transparency in THz frequency range) is very low around these zero-dispersion wavelengths, which is required for distortion less transmission of waves.

9.2 *Bandgap-Based Effective Indices and Dispersion in Photonic Bandgap Fibers (PBGFs)*

The photonic bandgap effect is used by Photonic Bandgap Fibers (PBGFs), in particular hollow-core fibers, to channel light within a core that may be hollow or have

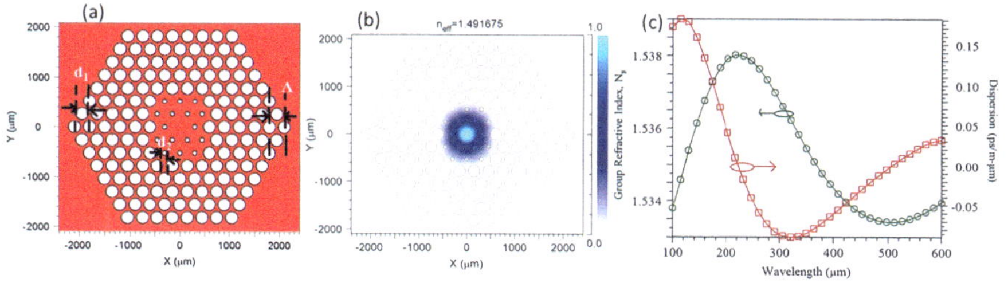

Fig. 9 **a** Schematic view of the special topology of air holes in the near core region of polyethylene solid core PCF. Here, the pitch, $\Lambda = 300$ μm, the large air holes, $d_1 = 0.8\Lambda$, and small air holes, $d_2 = 0.3\Lambda$. **b** Modal field profile for wavelength, $\lambda = 350$ μm. **c** Dispersion spectral characteristics (red) and group refractive index (green)

a lower refractive index than the surrounding cladding [30]. A photonic bandgap is produced by the cladding's periodic microstructure, which is usually made up of a lattice of air holes and prevents some light wavelengths from passing through it. Because of this phenomenon, light is forced to remain inside the core, giving rise to special guiding qualities. The bandgap characteristics of the cladding structure, which may be modified by changing the air holes' size, shape, and arrangement, define the effective indices of PBGFs. Because they affect how various light wavelengths travel through the fiber, these effective indices are essential in establishing the dispersion characteristics of the fiber. With this ability, dispersion may be controlled to an unprecedented degree, making it possible to construct fibers with customized dispersion profiles that are appropriate for a variety of uses, such as high-power laser delivery, low-loss transmission, and precise metrology.

Computational tool like the Finite Element Method [38] (FEM) are frequently employed to examine the intricate dispersion characteristics of hollow-core fibers and PBGFs. Finite Element Method (FEM) discretizes the fiber structure into a finite number of elements and solves Maxwell's equations within this discretized domain, enabling accurate modeling of the complex geometries of these fibers. This approach works especially well for managing the intricate boundary conditions and asymmetric geometries that are common to PBGFs. Researchers can precisely compute the fiber's effective indices and dispersion curves using FEM, accounting for the influence of the periodic cladding structure. The modal properties and loss characteristics of PBGFs may be thoroughly understood by FEM simulations, which make it possible to optimize fiber designs for uses. Using this method is crucial to creating fibers with customized dispersion characteristics, like flattened dispersion profiles or zero-dispersion wavelengths.

The Full Vector Plane Wave Expansion (FV-PWE) method is another advanced computational technique used to analyze the photonic bandgap properties of PBGFs and hollow-core fibers [39]. FV-PWE involves expanding the electromagnetic fields within the periodic cladding structure into a series of plane waves and solving the resulting eigenvalue problem to obtain the band structure of the fiber. This method provides a comprehensive understanding of the bandgap properties, including the

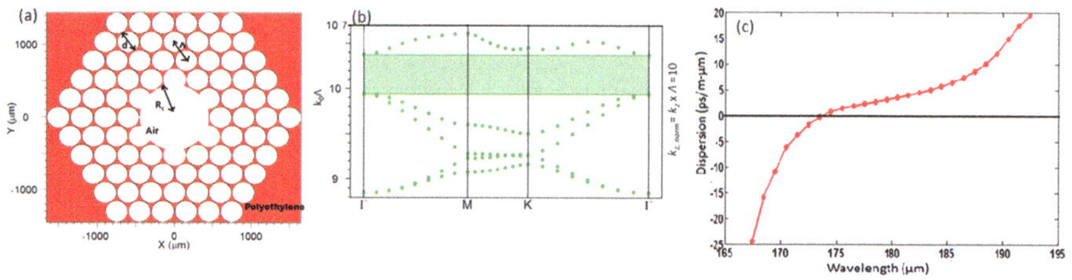

Fig. 10 **a** PBG fiber, **b** band gap and **c** spectrum of dispersion

frequency ranges where light propagation is prohibited and the corresponding effective indices. FV-PWE is particularly effective in capturing the vector nature of the electromagnetic fields, allowing for accurate modeling of polarization effects and higher-order modes.

We have designed a polyethylene photonic band gap fiber to guide the THz waves as described in Fig. 10a. The pitch, Λ is called distance between the centers of air holes which is considered as 300 μm. To achieve the photonic band gap for a range of THz wavelengths, the diameter of air holes, d should be very large. Here, in this design, we have considered d = 0.95 times pitch. The radius of the hollow core, R_c should also be large such that surface modes between the vacuum and polyethylene struts around the hollow core cannot couple to main fundamental mode inside the fiber. Figure 10b shows the photonic band gap which occurred for a particular normalized longitudinal wave vector (here normalized $k_{z,\,norm} = k_z \times \Lambda$ is considered as 10). For more illustration, please refer to [39]. This band gap has been calculated using full vector plane wave expansion method from RSoft module BandSOLVE. The dispersion (see Fig. 10c) has been calculated from the effective index versus wavelength curve. The effective index versus wavelength curve has been calculated using finite element method from RSoft module Femsim. The dispersion curve has one zero dispersion wavelength, around that the dispersion is very low.

9.3 Polarization Splitting Using Dual Core PCFs

Utilizing dual-core PCFs for polarization splitting allows for the effective separation of polarization states by taking advantage of the fiber's birefringent features. Two closely spaced cores are contained in a shared microstructured cladding, which in dual-core PCFs is usually made up of a periodic array of air holes (see our design for a THz polarization splitting dual core PCF in Fig. 11a). Different propagation constants for light polarized along different axes are produced by the geometry and arrangement of these cores. Researchers have also studied the arrangements of air holes inside the cores (as shown in Fig. 11b, c) such that effective indices versus frequency curves for different polarization modes behave such that splitting of polarization can be achieved. It is possible to produce significant coupling for one polarization state

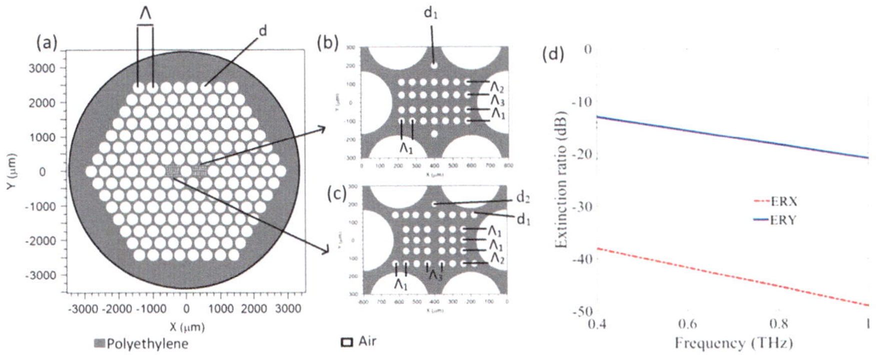

Fig. 11 **a** Schematic of dual core PCF, **b**, **c** air holes arrangement in the two cores and **d** extinction ratio for x and y polarized beams versus frequency

and preserve weak or no coupling for the orthogonal state by carefully arranging the separation and structure of the cores [40]. Effective polarization splitting results from this differential coupling, which allows light entering the fiber to be divided into its individual polarization components, each of which is primarily steered in one of the two cores [41]. For applications in polarization-sensitive devices, optical signal processing, and telecommunications, where maintaining and regulating polarization states is essential to system stability and performance this trait is extremely significant.

The right core receives the THz power input in both the x- and y-polarizations. The power in y-polarization couples fully from right core to left core at splitting length. In right core, the x-polarization's power is still present. For this polarization, very little power couples to right core. The ratio of desired to undesired polarization at a specific port or core is known as the extinction ratio. The extinction ratio of x and y-polarization for various terahertz frequencies have been calculated at proper splitting length. One can understand that the extinction ratios are high which demonstrates the high efficiency of polarization splitting.

9.4 THz Wave Generation Using Four Wave-Mixing in PCF

Terahertz (THz) sources can be produced by utilizing the nonlinear optical process of four-wave mixing (FWM) in photonic crystal fibers (PCFs) [33]. FWM involves the interaction of two or more optical waves in the highly nonlinear medium of the PCF, which is usually improved by the PCF's customized dispersion parameters and tight light confinement. The phase-matching conditions required for effective FWM can be obtained by carefully designing the microstructure of the PCF, which will allow pump photons to be converted into new frequencies. By choosing the right pump wavelengths and power levels and making sure the PCF's dispersion profile

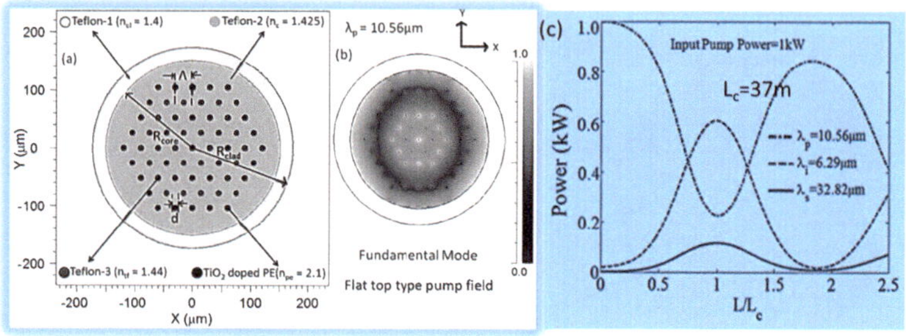

Fig. 12 **a** PCF with microstructured inclusions in the core, **b** the fundamental mode @ 10.56 μm wavelength and **c** the power evolution along the length of fiber

allows phase-matching over the required frequency range, this procedure can be utilized to produce THz radiation. PCFs offer special qualities that make them perfect for creating compact and effective THz sources, which are useful for spectroscopy, imaging, and communications. These traits include the capacity to control dispersion and improve nonlinear interactions. Figure 12a shows our designed PCF which has been used for FWM inside it to generate THz radiation. Figure 12b shows the length-based dependencies of pump, idler and THz signal power. At coherence length or so-called optimized length, the pump power has been decreased and this power has been gone to THz and idler power.

9.5 Supercontinuum Using PCF

Supercontinuum generation is a potent process in PCFs that converts a narrowband input pulse, usually from a femtosecond laser, into a continuous, incredibly wide spectrum that extends from the ultraviolet to the infrared. The distinct dispersion and nonlinear characteristics of PCFs are utilized in this procedure. PCFs' microstructured cladding makes it possible to precisely design dispersion, which is necessary for managing phase matching conditions and group velocity dispersion—two factors that are crucial for effective supercontinuum generation. Spectral widening is facilitated by nonlinear processes such self-phase modulation, four-wave mixing, and soliton dynamics, which are amplified by the close confinement of light in the PCF core [6, 7]. Supercontinuum light produced in PCFs is extremely coherent and useful in many different sectors, such as spectroscopy, metrology, optical coherence tomography, and telecommunications. It is a light source that is strong and adaptable in these domains.

We have designed diatomic Teflon PCF (see Fig. 13a) to achieve supercontinuum in THz wavelength regime (see Fig. 13b). Photonic crystal fibers (PCFs) use four-wave mixing (FWM), self-phase modulation (SPM), numerous nonlinear optical processes, including soliton production and cross-phase modulation (XPM)

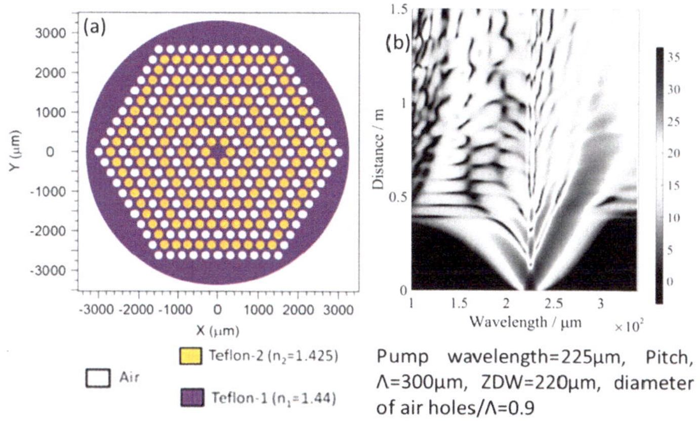

Fig. 13 **a** Diatomic PCF fiber with Teflon core, **b** supercontinuum achieved

to generate supercontinuum. Strong ultrafast laser pulses propagate through the PCF to create a supercontinuum, where solitons are formed by the interplay between dispersion and nonlinear processes.

Through their interaction, these solitons expand the spectrum and produce new frequencies through FWM and self-soliton Raman scattering. Furthermore, SPM broadens the spectrum range by modifying the pulse's phase during propagation, and XPM can further impact the spectral properties through interactions between various pulses or spectral components. The nonlinear Schrödinger equation (NLSE) [20, 21], which includes terms for the nonlinear response and dispersion, governs the evolution of these effects. It explains how the pulse's amplitude and phase change with distance in the nonlinear medium.

9.6 *Various Types of Sensing Using PCFs*

PCFs have special optical and structural characteristics that make them very flexible for use in sensing applications. Refractive index sensing is a form of sensing in which variations in the refractive index of the surrounding medium modify the properties of light propagation inside the fiber. PCFs are perfect for chemical and biological sensing since they may be used to detect changes in modal characteristics or shifts in transmission spectra. Sensitive materials can be coated or poured into the air pores of PCFs to improve sensitivity and the material's interaction with the target analyte [22, 23]. By functionalizing the surface of the fiber with certain receptors, PCFs can be used to feel biomolecules, monitor fluid refractive indices, and detect gas concentrations.

The sense of strain and temperature is another crucial use. High birefringence designs enable PCFs to be highly responsive to mechanical strain and temperature variations. When the fiber is exposed to changes in temperature or mechanical deformations, the microstructured cladding modifies the effective refractive index of the guided modes, resulting in detectable modifications in the transmission or reflection spectrum. Because of these characteristics, PCFs can be used in structural health monitoring, where they can identify temperature variations, stress, and strain in structures like bridges and buildings. Additionally, by utilizing its capacity to identify changes in pressure through variations in light transmission brought on by the compression of the fiber structure, PCFs can be employed for pressure sensing in a variety of industrial applications.

10 Conclusion

Because of their distinctive microstructured construction, PCFs give unmatched control over light propagation, marking a significant leap in the field of fiber optics. PCFs offer amazing dispersion tailoring, enhanced nonlinear interaction, and polarization state maintenance capabilities through precise engineering of the air hole patterns and core structures. PCFs are extremely ideal for a wide range of applications due to these qualities, which include effective polarization splitting and dispersion management, as well as the generation of supercontinuum light sources and THz radiation by four-wave mixing. Because of their flexibility, PCFs can also be tailored to meet particular needs, such enhancing sensitivity for sensing applications or decreasing losses for high-power transmission. The variety of sensing applications for PCFs emphasizes their significance and adaptability even further. Utilizing their sensitivity to temperature, strain, pressure, and refractive index variations, PCFs can be used in a variety of applications, including as industrial process control, chemical and biological sensing, and structural health monitoring. High sensitivity and specificity can be achieved by functionalizing the fiber surface with particular materials or coatings, which improves the interaction with target analytes. All things considered, the developments in PCF technology not only extend the capabilities of existing optical fibers but also create new opportunities for research and development in the fields of spectroscopy, metrology, sensor technologies, and telecommunications.

References

1. Ghatak, A.K., Thyagarajan, K.: Optical Electronics. Cambridge University Press (1989)
2. Ghatak, A.K., Thyagarajan, K.: An Introduction to Fiber Optics. Cambridge University Press (1998)
3. Snyder, W., Love, J.: Optical Waveguide Theory. Springer Science & Business Media (1983)
4. Snyder, W., Young, W.: Modes of optical waveguides. J. Opt. Soc. Am. **68**, 297–309 (1978)
5. Jacobs, F.B.: Optical fiber communication technology and system overview. In: Soares, O.D.D. (ed.) Trends in Optical Fiber Metrology and Standards, vol. 285, pp. 297–309. Springer, Dordrecht (1995)

6. Senior, J.M.: Optical Fiber Communications: Principles and Practice. Pearson Education India (2009)
7. McGloin, D., Simpson, N.B., Padgett, M.J.: Transfer of orbital angular momentum from a stressed fiber-optic waveguide to a light beam. Appl. Opt. **37**, 469–472 (1998)
8. Gibson, G., Courtial, J., Padgett, M.J., Vasnetsov, M., Pas'ko, V., Barnett, S.M., Franke-Arnold, S.: Free-space information transfer using light beams carrying orbital angular momentum. Opt. Express **12**, 5448–5456 (2004)
9. Bozinovic, N., Golowich, S., Kristensen, P., Ramachandran, S.: Control of orbital angular momentum of light with optical fibers. Opt. Lett. **37**, 2451–2453 (2012)
10. Morris, A.S., Langari, R.: Sensor technologies. In: Morris, A.S., Langari, R. (eds.) Measurement and Instrumentation, pp. 317–345. Butterworth-Heinemann (2012)
11. Kumar, A., Sharma, A.: An efficient nonparaxial and bidirectional wave propagation method for optical fibers & fiber components. Opt. Commun. **558**, 130375 (2024)
12. Kumar, A., Sharma, A.: Modal and propagation loss analysis of large mode area single-mode fibers. In: Frontiers in Optics + Laser Science 2023 (FiO, LS). Technical Digest Series, paper JM7A.43. Optica Publishing Group (2023)
13. Kumar, A., Sharma, A.: Propagation of higher order modes in FBGs. In: Frontiers in Optics + Laser Science 2022 (FiO, LS). Technical Digest Series, paper JTu5B.60. Optica Publishing Group (2022)
14. Kumar, R., Sharma, A.: A new method to analyze fiber Bragg gratings. Opt. Fiber Technol. **53**, 102017 (2019)
15. Bertoncini, A., Liberale, C.: 3D printed waveguides based on photonic crystal fiber designs for complex fiber-end photonic devices. Optica **7**, 1487–1494 (2020)
16. Broeng, J., Mogilevstev, D., Barkou, S.E., Bjarklev, A.: Photonic crystal fibers: a new class of optical waveguides. Opt. Fiber Technol. **5**, 305–330 (1999)
17. Hu, D.J.J., Ho, H.P.: Recent advances in plasmonic photonic crystal fibers: design, fabrication and applications. Adv. Opt. Photon. **9**, 257–314 (2017)
18. Kumar, V., Varshney, R.K., Kumar, S.: Near zero flat dispersion by hexagonal patterned unit cell in THz photonic crystal fiber. In: Springer Proceedings in Physics, vol. 258 (2019)
19. Kumar, V., Varshney, R.K., Kumar, S.: Near zero-dispersion broadband THz guidance in microstructured fibers. In: Proceedings of the International Conference of Fiber Optics and Photonics (Photonics-2018), paper no. SD1-C3 (2018)
20. Kumar, S., Kumar, V., Varshney, R.K.: Broad and flat supercontinuum generation (ranging from wavelengths 2.5 μm to 8 μm) using honeycomb microstructured fiber. J. Opt. **53**, 1202 (2024)
21. Kumar, V., Varshney, R.K., Kumar, S.: Terahertz supercontinuum in the biatomic photonic crystal fiber. In: Proceedings of the Conference on Lasers and Electro-Optics Europe & European Quantum Electronics Conference (CLEO/Europe-EQEC), poster session no. cc_p_32 (2019)
22. Kumar, V., Varshney, R.K.: Real-time monitoring of toxic CS_2 with high concentrations using a suitable terahertz photonic crystal fiber. J. Opt. Soc. Am. B **40**, 239–246 (2023)
23. Kumar, V., Varshney, R.K., Kumar, S.: Two interpenetrating hexagonal lattice in clad photonic crystal fiber for THz sensing. In: Proceedings of the Frontiers in Optics/Laser Science (FiO/LS), paper no. FM7B.5 (2020)
24. Debord, Amrani, F., Vincetti, L., Gérôme, F., Benabid, F.: Hollow-core fiber technology: the rising of 'gas photonics'. Fibers **7**, 16 (2019)
25. Jauregui, Limpert, J., Tünnermann, A.: High-power fibre lasers. Nat. Photon. **7**, 861–867 (2013)
26. Borzycki, K., Osuch, T.: Hollow-core optical fibers for telecommunications and data transmission. Appl. Sci. **13**, 10699 (2023)
27. Paixão, T., Araújo, F., Antunes, P.: Highly sensitive fiber optic temperature and strain sensor based on an intrinsic Fabry-Perot interferometer fabricated by a femtosecond laser. Opt. Lett. **44**, 4833–4836 (2019)
28. Taha, A., Ali, N., Sapiee, N.M., Fadhel, M.M., Mat Yeh, R.M., Bachok, N.N., Al Mashhadany, Y., Arsad, N.: Comprehensive review tapered optical fiber configurations for sensing application: trend and challenges. Biosensors **11**, 253 (2021)

29. Wang, Q., Farrell, G., Freir, T.: Theoretical and experimental investigations of macro-bend losses for standard single mode fibers. Opt. Express **13**, 4476–4484 (2005)
30. Kumar, V., Varshney, R.K., Kumar, S.: Broadband and flat THz dispersion in biatomic photonic band gap fibers. In: Proceedings of the Workshop on Recent Advances in Photonics (WRAP 2019), poster session, pp. 1–3 (2019)
31. Kumar, V., Varshney, R.K., Kumar, S.: Generation and beam steering of THz radiation in an actively modulated nonlinear crystal. Opt. Commun. **526**, 128950 (2023)
32. Ji, G., Huang, Z., He, W., Yin, R., Zheng, Y., Kumar, V., Jiang, X., Leng, Y., Pang, M.: Experimental studies on the core-structure dependence of backward Brillouin gain in solid-core photonic crystal fibers. Opt. Express **31**, 35742–35753 (2023)
33. Kumar, V., Varshney, R.K., Kumar, S.: Terahertz generation by four-wave mixing and guidance in diatomic Teflon photonic crystal fibers. Opt. Commun. **454**, 124460 (2020)
34. Arora, V., Kumar, V., Varshney, R.K.: Deformed core high birefringent photonic crystal fiber for THz guidance. In: Proceedings of the International Conference of Fiber Optics and Photonics (Photonics-2018), paper no. SP153 (2018)
35. Wang, Z., Zhao, C.L., Jin, S.: Design of a bending-insensitive single-mode photonic crystal fiber. Opt. Commun. **19**, 213–218 (2013)
36. Schreiber, T., Andersen, T.V., Schimpf, D., Limpert, J., Tünnermann, A.: Supercontinuum generation by femtosecond single and dual wavelength pumping in photonic crystal fibers with two zero dispersion wavelengths. Opt. Express **13**, 9556–9569 (2005)
37. Saitoh, K., Koshiba, M.: Full-vectorial imaginary-distance beam propagation method based on a finite element scheme: application to photonic crystal fibers. IEEE J. Quantum Electron. **38**, 927–933 (2002)
38. Koshiba, M.: Full-vector analysis of photonic crystal fibers using the finite element method. IEICE Trans. Electron. **85**, 881–888 (2002)
39. Shi, S., Chen, C., Prather, D.W.: Plane-wave expansion method for calculating band structure of photonic crystal slabs with perfectly matched layers. J. Opt. Soc. Am. A **21**, 1769–1775 (2004)
40. Kumar, V., Varshney, R.K., Kumar, S.: Design of a compact and broadband terahertz polarization splitter based on gradient dual-core photonic crystal fiber. Appl. Opt. **59**, 1974 (2020)
41. Kumar, V., Varshney, R.K., Kumar, S.: Compact and efficient polarization splitter based on dual core microstructured fiber for THz applications. Results Opt. **4**, 100094 (2021)

Smart Soft Materials: A Versatile Class of Materials for Frontier Technologies

Akansha Dixit, Dibyendu S. Bag, and Amodini Mishra

Abstract Smart Soft Materials (SSMs) represent a pioneering field at the intersection of material science and technology, characterized by their ability to respond to various external stimuli. This chapter provides a comprehensive overview of SSMs, organized into several key areas: responsive stimuli, composition, biomimetic materials, structural classification, applications, and functional properties. The chapter begins by detailing the different responsive stimuli that drive the behavior of SSMs, including thermo-responsive, photo-responsive, pH-responsive, magneto-responsive, electro-responsive, moisture-responsive, and mechanically responsive materials. It then categorizes SSMs based on their composition into polymeric smart materials (such as conductive, electroactive, and shape-memory polymers), hybrid smart materials (composite, multifunctional, and hybrid nanomaterials), and biomimetic smart materials inspired by natural systems. Further, the chapter explores the structural classification of SSMs, highlighting hydrogels, elastomers, liquid crystal elastomers (LCEs), and composite smart materials. It examines their diverse applications in biomedical fields, soft robotics, wearable technology, adaptive optics, and smart textiles. The discussion includes specific examples of how these materials are applied in real-world scenarios, showcasing their versatility and impact. The chapter concludes with an analysis of the functional properties of SSMs, such as self-healing, shape-memory, conductive, and stimuli-responsive behavior, underscoring their potential to revolutionize various technological domains. By integrating theoretical insights with practical examples, this chapter also provides a comprehensive overview of SSMs and their transformative role in modern technology.

A. Dixit (✉)
Department of Chemistry, Indian Institute of Technology Kanpur, Kanpur 208016, India
e-mail: akanshadixit17@gmail.com

D. S. Bag
Defence Materials and Stores Research and Development Establishment, DMSRDE PO, Kanpur 208013, India

A. Mishra
Department of Physics, Indian Institute of Technology, New Delhi 110016, India

A. Mishra and V. Pathak (eds.), *Smart Materials for Energy Storage and Biomedical Applications*, Engineering Materials, https://doi.org/10.1007/978-3-031-92584-9_7

Keywords Smart soft materials (SSMs) · Stimuli-responsive · Biomimetic materials · Polymeric smart materials · Hybrid smart materials · Functional properties

1 Introduction

Smart soft materials (SSMs) have emerged as a transformative class of materials that combine the inherent flexibility of soft materials with the ability to respond dynamically to external stimuli [1]. The physical and chemical properties of smart soft materials can change their properties or behavior in a controlled and reversible manner when exposed to different external stimuli such as light, solvents, mechanical strain, temperature, and electric or magnetic field [2]. These responsive features allow for the creation of various smart devices, including smart windows, dynamic displays, rewritable media, anticounterfeit tabs, encryption devices, stretchable electronics, wearable strain sensors, human–machine interfaces, thermal camouflage, and humidity sensors [3]. SSMs have emerged as a transformative class of materials that combine the inherent flexibility of soft materials with the ability to respond dynamically to external stimuli. These materials effectively bridge the gap between biology and technology by mimicking the flexibility, adaptability, and resilience found in natural systems. Unlike rigid materials, SSMs can easily conform to complex and dynamic surfaces, which allows them to interact more naturally and comfortably with their surroundings. These capability enhances their functionality in numerous applications, providing benefits such as improved performance in wearable devices, better integration in biomedical implants, and more efficient operation in soft robotics [4]. By adapting to changing environments and maintaining their performance, SSMs offer significant advantages over traditional materials, leading to more advanced and user-friendly technological solutions.

Their softness and tunable properties make SSMs ideal for applications requiring interaction with biological tissues or dynamic environments. In biomedical engineering, they enable biocompatible implants and prosthetics that adapt to body movements, enhancing patient outcomes. These materials have also revolutionized the fields of sensors, actuators, and robotics. Their ability to change properties in response to external stimuli enables the creation of highly sensitive and adaptable sensors, such as strain and pressure detectors, which are integral to wearable technology for monitoring health parameters. In actuators, materials like shape-memory polymers and liquid crystal elastomers allow for the development of soft actuators that can mimic natural muscle movements, enhancing the dexterity and functionality of soft robots [5]. These robots, made from smart soft materials, can navigate tight spaces, handle delicate objects, and adapt to dynamic environments, making them ideal for applications in healthcare, manufacturing, and exploration in various modern technologies. The integration of these materials in such technologies not only improves performance but also opens up new possibilities for advanced, adaptive systems across various industries. Smart soft materials can be classified based on various criteria, including their responsive stimuli, compositions, structures, and applications. One way to classify smart soft materials is by the type of stimuli responsive behaviour. Thermo-responsive materials change their properties with temperature, such as shape-memory polymers that return to their original shape when heated. Photo-responsive materials respond to light, exemplified by azobenzene-containing polymers that alter their structure under UV light [6]. pH-responsive materials, like certain hydrogels, adjust their properties with changes in pH [7]. Magneto-responsive materials, such as magnetorheological fluids, react to magnetic field, while electro-responsive materials change properties in response to electric field, like electrorheological fluids [8]. Moisture-responsive materials, such as hygroscopic polymers, respond to humidity changes, and mechanically responsive materials, like piezoelectric polymers, alter their properties under mechanical strain.

Classifying smart soft materials by composition reveals a range of types. Polymeric smart materials, including shape-memory and conductive polymers, exhibit smart behavior. Hybrid smart materials combine different material types, such as polymer-metal composites [9]. Biomimetic smart materials, inspired by natural materials like silk, mimicking biological functions and structures. Considering their structures, smart soft materials include hydrogels, which are crosslinked polymer networks capable of holding large amounts of water, polyacrylamide (polyAM), poly(vinyl alcohol) (PVA) and cellulose hydrogels [10]. Elastomers are highly elastic materials that return to their original shape after deformation, such as silicone rubber. Liquid crystal elastomers (LCEs) combine properties of liquid crystals and elastomers, seen in nematic LCEs [11]. Composite smart materials consist of two or more distinct phases, like polymer-matrix composites.

SSMs can also be classified based on their applications. In biomedical fields, these materials are used in drug delivery, tissue engineering, and medical implants, with hydrogels and shape-memory polymers being prominent [12]. In soft robotics, materials like liquid crystal elastomers and dielectric elastomers are used in actuators, sensors, and flexible robotic components. Wearable technology is benefited from conductive polymers and stretchable sensors for health monitoring and flexible electronics. Adaptive optics use materials like liquid crystal-based smart windows for lenses and other optical components that can change properties. Smart textiles integrate materials like shape-memory fibres into fabrics for adaptive clothing and responsive materials.

SSMs classification based on different criteria easy to understand their diverse functionalities and potential applications across various fields showing in Fig. 1. For instance, categorizing them by responsive stimuli reveals, these materials can adapt to changes in their environment [13]. Similarly, examining their composition, structure, application, and functional properties provides insights into their unique characteristics and they can be effectively utilized in innovative technologies. This detailed classification approach makes it easier to identify the specific advantages of each type of smart soft material, facilitating their development and integration into practical applications in areas like biomedical engineering, soft robotics, wearable technology, adaptive optics, and smart textiles [14].

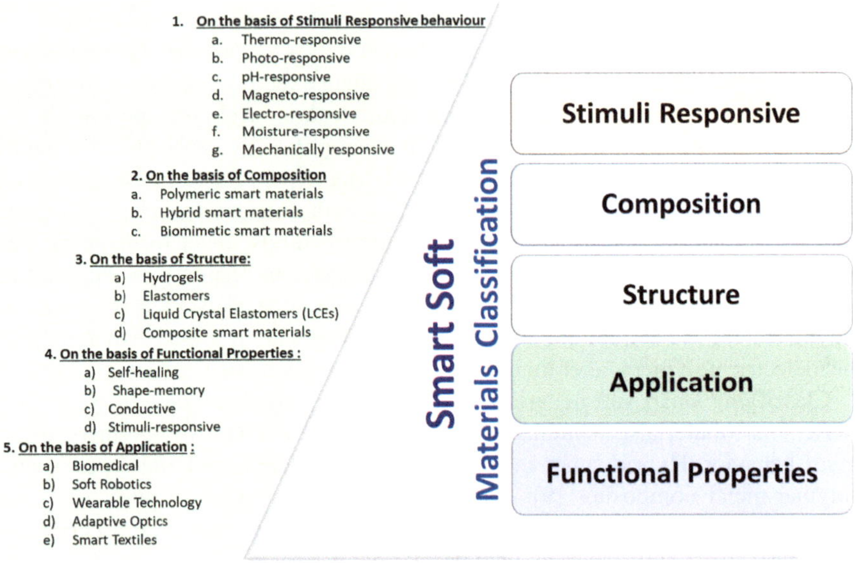

Fig. 1 Classification of smart soft materials: a comprehensive overview. The classification was based on five key factors, categorized into: (1) stimuli responsiveness, (2) composition, (3) structure, (4) application, and (5) functional properties. Each category is further detailed with specific subgroups to highlight their distinct features and attributes

2 Classification of Smart Soft Materials

Smart soft materials (SSMs) are a versatile class of materials that have gained significant attention due to their ability to adapt and respond to various external stimuli. These materials can be classified based on several criteria, including their responsive stimuli, composition, structure, applications, and functional properties [15]. By understanding these classifications, researchers and engineers can better harness the unique capabilities of SSMs for a wide range of innovative technologies. Each classification highlights the diverse functionalities and potential applications of these materials, making them indispensable in advancing modern technology. These classifications help in identifying the most suitable materials for specific applications, such as biomedical devices, soft robotics, wearable technology, adaptive optics, and smart textiles. Additionally, functional properties like self-healing, shape-memory, and conductivity further define the utility of smart soft materials in various fields. This multifaceted classification approach provides a comprehensive understanding of SSMs, paving the way for their integration into next-generation technological solutions.

3 Smart Soft Materials and Their Stimuli-Responsive Behavior

Stimuli-responsive materials are a class of advanced materials engineered to undergo measurable and reversible changes in any of their properties in response to external stimuli. These materials are designed to dynamically adapt their properties in real time, making them highly versatile for a range of applications including sensors, actuators, and drug delivery systems [16]. By leveraging their responsiveness, these materials can be tailored to perform specific functions or exhibit desirable characteristics under varying conditions, which significantly enhances their functionality and efficiency in practical applications (Table 1).

The classification of stimuli-responsive materials is generally based on the type of stimulus they respond to and the nature of their response. This includes thermo-responsive materials that change properties with temperature fluctuations, pH-responsive materials that respond to changes in acidity or basicity, and photo-responsive materials that alter their behavior under light exposure and many others. Each classification involves distinct mechanisms and material compositions, such

Table 1 Comprehensive classification and examples of smart soft materials based on responsive stimuli

External stimuli	Examples	Application
Thermo-responsive	1. Poly(N-isopropylacrylamide) (PNIPAM) [17] 2. Oligo(ethylene glycol-methacrylate) based polymers [18]	Used in drug delivery and tissue engineering
Photo-responsive	1. Azobenzene-containing polymers [19] 2. Spiropyran-based hydrogels [20]	Used in optical data storage and sensors
pH-responsive	1. Poly(acrylic acid) (PAA) [21] 2. Chitosan-based hydrogels [22]	Used in drug delivery and waste treatment
Magneto-responsive	1. Magnetite nanoparticles embedded in polymers [23] 2. Iron oxide nanoparticle composites [24]	Used in targeted drug delivery and actuators
Electro-responsive	1. Polyvinylidene fluoride (PVDF) [25] 2. Dielectric elastomers [26]	Used in sensors and actuators
Moisture-responsive	1. Cellulose-based hydrogels [27] 2. Polyvinyl alcohol (PVA) [28]	Used in hygiene products and sensors
Mechanically responsive	1. Polyurethane shape-memory polymers [29] 2. Silly Putty (polydimethylsiloxane) [30]	Used in soft robotics and medical devices

as hydrogels, liquid crystals, and conductive polymers, that define their applications and performance [31]. Understanding these classifications is crucial for developing advanced technologies and innovative solutions in various fields ranging from biomedical engineering to environmental monitoring.

3.1 Temperature Responsive Smart Soft Materials

Temperature-responsive SSMs are engineered to exhibit significant changes in their physical or chemical properties in response to temperature variation. These materials are characterized by their ability to undergo reversible phase transitions, which make them highly versatile for various applications in different fields such as biomedical engineering, environmental sensing, and smart textiles. This section delves into the mechanisms behind temperature responsiveness, the different types of materials used, and their practical applications [32].

Mechanisms of Temperature Responsiveness: These materials typically function through one of several mechanisms, including phase transitions, changes in polymer chain conformation and molecular interactions. A common mechanism involves the temperature-induced phase transition of the material which is shown in Fig. 2a. For example, certain hydrogels exhibit a lower critical solution temperature (LCST) or an upper critical solution temperature (UCST), where the material undergoes a transition between hydrophilic and hydrophobic states depending on the critical temperature. At temperatures below the LCST, these hydrogels swell due to their affinity for water molecules due to formation of H-bonding, while above the LCST, they collapse and expel water, leading to a significant change in volume [31]. Another mechanism involves the change in polymer chain conformations. Thermo-responsive polymers, such as poly(N-isopropylacrylamide) (PNIPAM) [33], polyethylene glycol (PEG)-block-PNIPAM [34], and poly(N-vinylcaprolactam) (PNVCL) [35], exhibit significant conformational changes at their respective lower critical solution temperatures (LCST). For PNIPAM, below the LCST, the polymer remains in a hydrated, expanded state, while above this temperature, the chains collapse into a more compact, hydrophobic form. Similarly, PEG transitions from a hydrated, soluble state below was critical temperature to a more aggregated, hydrophobic state above this temperature, affecting their interaction with water. PNVCL is also changed from a hydrophilic, expanded state below its LCST to a hydrophobic, collapsed form above its LCST, altering its solubility and physical structure in response to temperature variation.

These materials have a wide range of practical applications, as biomedical field, they are employed in controlled drug delivery systems, tissue engineering scaffolds, and as sensors for monitoring body temperature. For example, smart hydrogels can release therapeutic agents at specific body temperatures, providing localized treatment for conditions such as arthritis or cancer. In environmental monitoring, thermo-responsive materials are used for temperature sensing and as part of adaptive

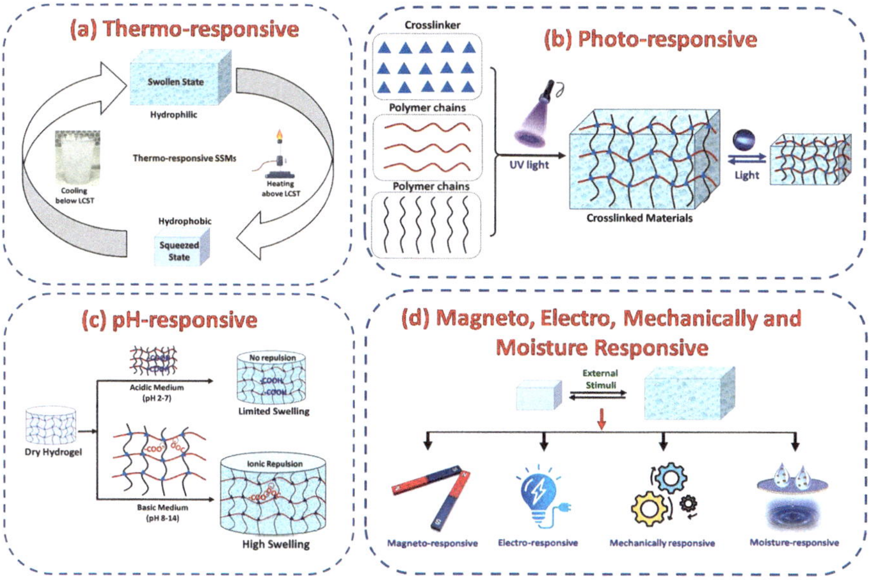

Fig. 2 Schematic illustration of different type of stimuli-responsive behavior of smart soft materials. **a** Thermo-responsive SSMs showing a reversible transformation between two states in response to heat. **b** Polymers and crosslinker in the presence of UV light exposure, crosslinked network formed and that material shows the responsive behavior in different stimuli. **c** Schematic representation of the pH-responsive phenomenon in the swelling under different pH conditions due to ionic attraction and repulsion. **d** SSMs materials response different stimuli like magneto, electro, mechanically and moisture responsive materials

systems that respond to environmental changes. These materials can be integrated into smart textiles to create garments that adapt to body temperature, enhancing comfort and performance in varying conditions [14]. Their unique mechanisms and diverse applications underscore their importance in advancing technology and improving everyday life. Continued research and development in this field will likely yield new materials and technologies that harness the power of temperature responsiveness in innovative ways. Innovations in synthesis and fabrication techniques will also play a crucial role for expanding the range of applications for temperature-responsive materials.

3.2 *Photo-Responsive Smart Soft Materials*

Photo-responsive SSMs are engineered to undergo reversible changes in their physical or chemical properties in response to light exposure. These materials are of significant interest due to their ability to respond dynamically to optical stimuli, offering a range of applications in the fields such as adaptive optics, imaging technologies, and smart surfaces [36].

Mechanisms of Photo-Responsiveness: The fundamental mechanism behind photo-responsive materials involves the absorption of photons, which initiates a photo-chemical or photo-physical reaction within the materials. This reaction typically results in changes to the materials' structures, properties, or interactions with the environment. Key mechanisms include photochromism, where the material alters it's colour in response to light exposure which is shown in Fig. 2b, and photoisomerization, where the material undergoes a reversible change in molecular structure. (a) *Photochromism*: These materials change their colour or transparency in response to light, typically through reversible chemical reactions for example, include spiropyrans and azobenzene, which convert between colourless and coloured forms upon exposure to UV or visible light. Other notable photo-responsive compounds are diarylethene [37], which can switch between open and closed ring forms, and fulgides, which exhibit reversible colour changes. The changes occur due to the formation of different chemical structures that absorb light differently. Photochromic materials are widely used in applications such as adaptive eyewear, dynamic optical filters, and light-sensitive coatings. (b) *Photoisomerization*: These materials upon light exposure induces a change in the molecular structure, resulting in significant alterations in physical properties. For instance, azobenzene derivatives undergo reversible trans–cis isomerization when exposed to UV or visible light, leading to changes in the material's shape, mechanical properties, or solubility. Another example is stilbenes, which can isomerize between different structural forms, affecting their optical and electronic properties. Ruthenium-based compounds also exhibit photoisomerization, where light-induced structural changes can impact their catalytic activity and coordination properties [38]. Additionally, spiropyrans and spirooxazines undergo photo-induced isomerization that alters their colour and physical properties, making them useful in applications such as smart coatings and light-responsive materials [39]. These photoisomerizable materials find applications in light-driven actuators, smart coatings, and responsive surfaces, where their ability to change properties under light stimuli is leveraged.

This class of SSMs have a broad range of applications due to their ability to dynamically respond to light. In the field of adaptive optics, these materials are used to create adjustable lenses and filters that can be tuned in real-time by controlling light exposure. In imaging technologies, these are used in devices such as optical switches and modulators. In biomedical applications they are utilized in drug delivery systems where light can be used to trigger the release of therapeutic agents at targeted locations and in tissue engineering, these materials can be used to create scaffolds that change their mechanical properties or bioactivity in response to light, facilitating controlled tissue growth and repair [40]. Future research will focus on enhancing their responsiveness, stability, and functionality. Innovations in material synthesis and design aim to improve their efficiency and expand their applications in adaptive systems and smart devices. These advancements could lead to breakthroughs in responsive coatings and new technologies, emphasizing the importance of photo-responsive materials in technological innovation.

3.3 pH-Responsive Smart Soft Materials

pH-responsive smart soft materials are designed to undergo reversible changes in their physical or chemical properties in response to variations in pH of the media. These materials are particularly valuable in the fields such as biomedical engineering, environmental sensing, and controlled drug delivery due to their ability to respond to the acidity or alkalinity of their environment.

Mechanisms of pH Responsiveness: The core mechanism behind pH-responsive materials involves changes in the ionization state of specific functional groups within the material in response to pH fluctuations (Fig. 2c). These materials often contain acidic or basic groups that gain or lose protons, leading to alterations in their solubility, swelling behavior, or mechanical properties. pH-responsive materials typically include functional groups such as carboxyl, amino, or phosphate groups that undergo ionization or de-ionization in response to pH change [21]. For example, carboxyl groups become negatively charged in basic conditions and neutral in acidic environments, affecting the materials' solubility and swelling properties. The swelling and shrinking behavior were shown by many pH-responsive materials that exhibit significant changes in volume or swelling behavior depending on the pH of the surrounding environment. Hydrogels containing pH-sensitive polymers can swell or shrink as pH changes, enabling controlled release of encapsulated substances or alteration in mechanical properties. pH-responsive materials include hydrogels, polymers, and nanoparticles that respond to changes in pH. Hydrogels like poly(acrylic acid) (polyAA) swell in alkaline conditions and collapse in acidic environments, making them ideal for controlled drug delivery. pH-sensitive polymers, such as poly(2-hydroxyethyl methacrylate) (polyHEMA), alter their hydrophilicity based on pH, which is useful in tissue engineering and pH sensing [41].

pH-responsive materials have a wide range of applications, particularly in controlled drug delivery, tissue engineering, and environmental sensing. In drug delivery systems, these materials can be used to release drugs at specific pH conditions, such as in the acidic environment of the stomach or the more alkaline conditions of the intestines. This controlled release capability enhances therapeutic efficacy and minimizes side effects. In tissue engineering, they can create scaffolds that respond to the pH of the surrounding environment, promoting tissue growth or repair. For instance, pH-sensitive hydrogels can be used as scaffolds that swell or shrink to facilitate the integration of tissue or to release growth factors in response to changes in the local pH. For environmental applications, these materials can be used for sensing and remediation. pH-sensitive sensors can detect changes in environmental conditions, while materials that alter their properties in response to pH can also be used to remove pollutants or contaminants from water. pH-responsive smart soft materials offer significant potential due to their ability to dynamically respond to changes in pH [42]. Their diverse mechanisms and applications underscore their importance in advancing technology and developing innovative solutions across various industries. Advances in material synthesis and design will aim to create materials that can respond to a broader range of pH values and exhibit more precise control over their properties. Innovations in this field could lead to new applications in biomedical devices, environmental monitoring, and beyond, further expanding the impact of pH-responsive materials.

3.4 Magneto-Responsive Smart Soft Materials

Magneto-responsive SSMs are engineered to undergo reversible changes in their physical or chemical properties in response to an external magnetic field. These materials are gaining significant attention for their potential applications in fields such as biomedical engineering, environmental sensing, and advanced manufacturing [43]. *Mechanisms of Magneto-Responsiveness:* The fundamental mechanism behind magneto-responsive materials involves the interaction of magnetic field with embedded magnetic particles within the materials. When exposed to a magnetic field, these materials experience forces that induce changes in their physical properties, such as shape, mechanical strength, or alignment (Fig. 2d). (i) Magnetic Particle Interaction: these materials are incorporated magnetic particles, such as iron oxide nanoparticles, which interact with external magnetic field. The magnetic particles can align or move within the material, leading to changes in its properties [33]. For example, magnetic field can induce deformation or alteration in mechanical properties of the material. Additionally, magnetic particles can affect thermal conductivity, as seen in magnetic composites that adjust heat transfer. Magnetic hydrogels can swell or shrink in response to the field, while magnetically responsive capsules can release substances based on magnetic stimulation. (ii) Magneto-Responsive Composites: composites incorporating magnetic particles with matrices like elastomers or foams exhibit unique magneto-responsive properties. For instance, magneto-responsive foams can adjust their shock absorption and impact resistance when exposed to magnetic field [44]. These can serve in smart actuators, allowing precise control of movement and flexibility. Additionally, magnetic field can modify thermal conductivity in adaptive insulation materials, optimizing heat management in various applications.

These materials have a broad range of applications due to their ability to respond dynamically to the magnetic field. In biomedical engineering, these materials are used in targeted drug delivery systems, where magnetic field can control the release of drugs, magnetic nanoparticles to specific locations within the body. These materials are also utilized in medical imaging and diagnostics, where magnetic particles can enhance contrast provide real-time imaging capabilities.

3.5 *Electro-Responsive Smart Soft Materials*

Electro-responsive SSMs dynamically alter their properties in response to electric field, enabling diverse applications. Electroactive polymers (EAPs), such as poly(pyrrole) and poly(3,4-ethylenedioxythiophene) (PEDOT), change their shape or volume due to electrochemical reactions induced by an electric field, making them valuable for actuators and artificial muscles [45]. Similarly, electro-responsive hydrogels, like poly(acrylamide) integrated with conductive polymers or nanoparticles, adjust their swelling behavior in response to electric field, which is beneficial for applications in drug delivery systems and tissue engineering [26]. These materials exemplify how electric fields can be harnessed to control and manipulate physical properties for advanced technological solutions. These materials include electroactive polymers, hydrogels, and actuators, each responding uniquely to electric fields. Polymers such as poly(pyrrole) and PEDOT undergo significant changes in volume or shape, making them ideal for applications in soft robotics and adaptive structures where precise control is needed [45]. Electro-responsive hydrogels, like poly(acrylamide) with conductive nanoparticles, adjust their swelling properties in response to electric field, facilitating controlled drug delivery [25]. Electro-responsive actuators, including dielectric elastomers, alter their shape or mechanical properties when exposed to electric field, which is beneficial for robotics and adaptive optics applications requiring precise and reversible changes [46].

Electro-responsive materials are utilized in diverse applications owing to their dynamic property changes in response to electric field. In soft robotics, electroactive polymers and actuators facilitate the development of flexible, responsive robots that can perform intricate movements. In biomedical engineering, electro-responsive hydrogels and polymers are employed in controlled drug delivery systems and tissue engineering scaffolds, offering precise control over drug release and scaffold behavior [46]. Additionally, these materials are integral to adaptive optics and electronic devices, where their variable properties support advanced functionalities.

3.6 *Moisture-Responsive Smart Soft Materials*

Moisture-responsive SSMs are engineered to exhibit reversible changes in their properties when exposed to variations in moisture levels. These materials demonstrate dynamic behaviour such as swelling, shrinking, or altering their mechanical properties in response to changes in humidity or moisture content [27]. This responsiveness makes them valuable materials across a range of applications, including biomedical engineering, environmental sensing, and adaptive textiles.

Moisture-responsive materials exhibit various mechanisms of responsiveness when exposed to different moisture levels. (i) Swelling and shrinking: are common phenomena, where materials like hydrogels and superabsorbent polymers undergo significant volumetric changes; for instance, poly(acrylic acid) hydrogels swell in high humidity and contract in drier conditions, which is useful for controlled drug delivery systems [28]. (ii) Mechanical property alterations: occur when materials, such as moisture-responsive elastomers, change in stiffness or elasticity with moisture exposure, enhancing their use in adaptive textiles and smart coatings. (iii) Chemical and structural changes: can happen as moisture induced reversible chemical reactions, affecting surface properties or structural integrity, which is leveraged for various applications.

Moisture-responsive materials will focus on improving their responsiveness, durability, and versatility. Advances in material design and synthesis will aim to create materials with enhanced performance across a wide range of moisture conditions [47]. Innovations could lead to new applications and improved functionalities in existing domains, expanding the impact of moisture-responsive smart soft materials.

3.7 Mechanical Responsive Smart Soft Materials

Mechanical responsive SSMs are designed to exhibit dynamic changes in their physical properties in response to mechanical stimuli such as stress, strain, or pressure [48]. These materials adapt their behavior based on mechanical inputs, enabling a wide range of applications from self-healing materials to adaptive structural components.

These materials exhibit diverse mechanisms that enable them to adapt dynamically to mechanical stimuli. (i) Shape memory effect: shape memory polymers (SMPs) and shape memory alloys (SMAs) can revert to a predefined shape after deformation when exposed to specific stimuli such as temperature. For instance, SMPs like polyurethanes or polycaprolactone, deformed at a lower temperature, return to their original form upon heating, making them suitable for applications in biomedical devices and deployable structures that require shape transitions in response to thermal changes [29]. (ii) Self-healing: materials represent another key mechanism, incorporating embedded microcapsules or reversible chemical bonds that facilitate autonomous repair of damage. Polymeric materials with these features can mend cracks and restore functionality after mechanical damage, significantly extending the lifespan of materials used in infrastructure, coatings, and electronics. (iii) Responsive textures: these materials that alter their surface properties, such as roughness or smoothness, in response to mechanical forces. For example, stimuli-responsive surfaces can become rougher or smoother under pressure or stretching, which enhances their functionality in applications like grippers or dynamic surfaces used in robotics and adaptive optics [30]. These mechanisms show the versatility and functionality of mechanical responsive materials, driving advancements across various technological domains and to develop materials with enhanced mechanical properties and multifunctional capabilities, broadening their applications and effectiveness in various fields. Continued advancements will drive the evolution of smart materials, leading to new possibilities in adaptive and self-repairing technologies.

4 Smart Soft Materials and Their Compositions

SSMs can be classified based on their composition into several categories, each with unique properties and applications (Table 2).

4.1 Polymeric Smart Soft Materials

These materials combine the inherent flexibility and lightweight nature of polymers with functional properties that make them ideal for a wide range of applications, including biomedical devices, soft robotics, sensors, and actuators [68]. Polymeric smart materials encompass a diverse range of materials with unique properties and applications. Such as:

(i) *Conductive Polymers*, such as polyaniline (PANI), polypyrene (PPy), and poly(3,4-ethylenedioxythiophene) (PEDOT), combine the electrical conductivity of metals with the flexibility of traditional polymers, making them suitable for flexible electronics, wearable sensors, and artificial muscles due to their ability to change electrical properties under mechanical deformation [15].

(ii) *Electroactive Polymers (EAPs)*, which include ionic EAPs like ionic polymer-metal composites (IPMCs) and electronic EAPs like dielectric elastomers, change shape or size in response to electric fields, making them valuable materials for actuators and haptic devices due to their significant strain capabilities and rapid response time [69].

(iii) *Shape Memory Polymers (SMPs)* exhibit a shape memory effect, returning to their original shape from a deformed state upon exposure to stimuli such as heat, which is due to reversible phase transitions, and are employed in medical devices, self-healing materials, and deployable aerospace structures [44].

Such SSMs were governed by several mechanisms, say, electrochemical reactions in conductive polymers involve redox processes that change their electrical properties. Electrostatic forces cause EAPs to deform in response to electric fields. Thermal transitions in Shape Memory Polymers (SMPs) involve phase changes, such as glass transition or melting, that enable them to return to their original shape. Swelling and deswelling in hydrogels and other stimuli-responsive polymers occur when they absorb or expel water, altering their volume and mechanical properties in response to environmental changes. These mechanisms enable polymeric smart materials to adapt and respond dynamically to various stimuli. Polymeric smart materials are promising, with ongoing advancements aimed at improving their responsiveness, durability, and functionality. Innovations in nanotechnology, such as integrating nanomaterials like graphene and carbon nanotubes, are expected to enhance their properties and performance. Additionally, developing biocompatible and biodegradable polymers will broaden their applications in medical and environmental fields.

Table 2 Classification and examples of smart soft materials based on composition

SSMS on the basis of composition	Polymeric SSMs	Description	Examples
Polymeric SSMs	Conductive polymers	Polymers that conduct electricity	1. Polyaniline (PANI) [49] 2. Polypyrrole (PPy) [50] 3. Poly(3,4-thylenedioxy thiophene) (PEDOT) [45] 4. Polyacetylene [51]
	Electroactive polymers	Polymers that change shape or size when stimulated by an electric field	1. Polyvinylidene fluoride (PVDF) [25] 2. Polyaniline-based actuators [49] 3. Ionic polymer-metal composites (IPMC) [24] 4. Dielectric elastomers [25]
	Shape memory polymers	Polymers that return to a predefined shape when exposed to the appropriate stimulus	1. Polyurethane (PU) [29] 2. Polycaprolactone (PCL) [52] 3. Polystyrene (PS) [53] 4. Epoxy-based shape memory polymers [54]
Hybrid SSMs	Composite hybrid	Materials that combine polymers with other materials like metals or ceramics to enhance their properties	1. Polymer-ceramic nanocomposites [55] 2. Carbon fiber-reinforced polymers [56] 3. Glass fiber-reinforced polymers [57] 4. Metal-polymer hybrids [58]
	Multi-functional hybrid	Materials designed to perform multiple functions simultaneously, such as structural support and self-healing	1. Self-healing composites [44] 2. Conductive and magnetic hybrid materials [31] 3. Thermal and electrical conductive hybrids [59] 4. Flame retardant hybrids [60]

(continued)

Table 2 (continued)

SSMS on the basis of composition	Polymeric SSMs	Description	Examples
	Hybrid nano-materials	Nanoscale materials that combine different types of substances to achieve unique properties	1. Graphene oxide/polymer composites [61] 2. Carbon nanotube/polymer composites [62] 3. Metal–organic framework (MOF) polymers [63] 4. Nanoclay/polymer composites [61]
Biomimetic SSMs		Materials inspired by natural systems that mimic biological functions or processes	1. Synthetic spider silk [64] 2. Bio-inspired hydrogels mimicking cartilage [65] 3. Mussel-inspired adhesives [66] 4. Lotus leaf-inspired superhydrophobic surfaces [67]

This table categorizes smart soft materials into polymeric, hybrid, and biomimetic types, providing specific examples for each sub-category

4.2 Hybrid Smart Soft Materials

Hybrid smart materials combine different types of integrated traditional materials with advanced functionalities to achieve enhanced performance and versatility. By merging the properties of various materials, hybrid systems aim to leverage the strength of each component, resulting in innovative solutions for complex applications [55]. Hybrid smart materials are classified based on their composition and the integration of multiple functionalities.

(i) ***Composite Hybrids*** blend smart polymers with traditional materials like metals or ceramics to enhance properties such as mechanical strength, electrical conductivity, or thermal stability. For instance, combining conductive polymers with conventional materials results in composites with improved flexibility and conductivity, making them ideal for use in flexible electronics and sensors [53]. These hybrids take advantage of the inherent properties of both components to create materials that are both robust and versatile. For example, combining conductive polymers like polyaniline (PANI) with carbon fibers results in composites that exhibit both high electrical conductivity and mechanical strength [49]. These composites are used in applications such as flexible electronics, wearable sensors, and lightweight structural components in aerospace engineering.

(ii) ***Multifunctional Hybrids*** incorporate multiple smart functionalities into a single material. The integration of shape memory polymers (SMPs) with conductive materials to develop hybrids that respond to both thermal stimuli and electrical signals [59]. This combination enables advanced applications such as self-healing structures and sophisticated actuators, where the materials' ability to change shape and conduct electricity provides enhanced control and adaptability. An example is composite of shape memory polymers (SMPs) with Graphene Oxide Containing Phosphorus [60]. This hybrid can change its shape in response to temperature change and conduct electricity. It was used in applications like smart actuators and self-healing materials.

(iii) ***Hybrid Nanomaterials*** involve embedding nanoparticles like carbon nanotubes or graphene into polymer matrices, significantly boosting mechanical strength, electrical conductivity, and thermal stability [62]. For Example, incorporating of graphene into polymers like poly(vinyl alcohol) (PVA) creates a hybrid material with enhanced mechanical strength, electrical conductivity, and thermal stability. These nanocomposites are used in high-performance sensors, flexible electronics, and advanced coatings.

Hybrid smart materials represent a significant advancement in material science, offering a combination of functionalities that surpass those of individual smart materials. By integrating various materials and functionalities, hybrid systems provide innovative solutions to complex problems across multiple industries [63]. Advances in nanotechnology, materials science, and manufacturing techniques are expected to lead to the creation of more sophisticated hybrid systems with improved properties. Hybrid smart materials will continue to play a crucial role in advancing technology and addressing emerging challenges in the fields ranging from biomedical engineering to electronics and environmental science.

4.3 Biomimetic Smart Soft Materials

These materials are designed to replicate the functionality and performance of biological systems and organisms. By emulating natural processes and structures, these materials achieve advanced capabilities that often surpass traditional engineering solutions. These materials draw inspiration from nature to address complex challenges in various fields, including robotics, medicine, and environmental technologies [65]. One of the most prominent examples of biomimetic smart materials was the lotus-leaf-inspired superhydrophobic surface. The lotus leaf is renowned for its remarkable water-repellent properties, which are attributed to their micro- and nanoscale surface structure, that was the structure causes water droplets to bead up and roll off, carrying dirt and contaminants with them a phenomenon known as the "lotus effect." For instance, a synthetic superhydrophobic coating is developed using a combination of nanostructured polystyrene and fluorinated polymers to mimic the lotus leaf's surface roughness and chemical properties. The resulting material exhibits exceptional water repellence, self-cleaning properties, and resistance to contamination [67].

The super hydrophobicity of these materials is achieved through a combination of surface roughness and low surface energy. The engineered surface texture mimics the hierarchical structure of the lotus leaf, while the use of fluorinated polymers reduces the surface energy, preventing water from spreading. As a result, water droplets form nearly spherical shapes and roll off the surface with ease, removing dirt and debris. The lotus-leaf-inspired superhydrophobic surfaces have a wide range of applications. In self-cleaning coatings, these materials are used to create surfaces that remain clean with minimal maintenance, such as in windows, facades, and solar panels. In medical devices, superhydrophobic surfaces help reduce bacterial adhesion, thereby improving the hygiene and longevity of implants and prosthetics. Additionally, they are used in textiles to produce water-resistant and stain-resistant fabrics.

5 Smart Soft Materials and Their Structures

SSMs are increasingly recognized for their ability to respond adaptively to external stimuli, and their classification based on structure provides a framework for understanding their diverse functionalities and applications. Mainly there are four major structural categories of smart soft materials: hydrogels, elastomers, liquid crystal elastomers (LCEs), and composite smart materials. Each category is defined by its unique structural characteristics, which contribute to its specific responsive behaviour and applications.

5.1 Hydrogels

Hydrogels are three-dimensional (3D), cross-linked polymer networks that can absorb and retain large amounts of water while maintaining their structural integrity showing in Fig. 3. Their responsiveness is primarily driven by changes in environmental conditions such as pH, temperature, or ionic strength of the medium. For instance, *temperature-responsive hydrogels* like poly(N-isopropylacrylamide), poly(NIPAM) exhibit a lower critical solution temperature (LCST), causing them to swell or shrink in response to temperature change below and above LCST respectively [70]. pH-responsive hydrogels, such as poly(acrylic acid), exhibit less swelling at low pH levels and increased swelling at high pH levels [32]. This behavior is due to the ionization of carboxyl groups within the polymer network, which causes the hydrogel to expand in alkaline conditions due to ionic repulsion. These property

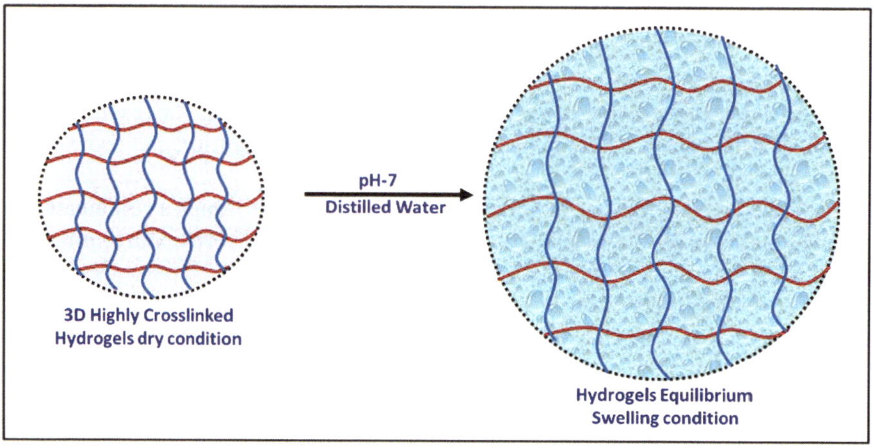

Fig. 3 Schematic illustration of hydrogels (3D-crosslinked network) in dry state and swollen hydrogels

makes hydrogels suitable for applications in drug delivery systems, where they can release therapeutic agents in response to specific stimuli, and in tissue engineering, where they provide a hydrated environment conducive to cell growth.

5.2 Elastomers

These are polymer materials characterized by their high elasticity and ability to return to their original shape after deformation. In the context of smart soft materials, elastomers can be engineered to respond to external stimuli such as electric field, magnetic field, or temperature change. Electroactive polymers (EAPs), such as dielectric elastomers, fall into this category. They deform significantly under an electric field, enabling their use in flexible actuators and artificial muscles. An example is *VHB 4910*, a commercially available dielectric elastomer commonly used in the construction of actuators and sensors due to its excellent deformability and energy density [71]. These materials are valued for their large strain capabilities, high energy density, and fast response times, making them ideal for applications in robotics, haptic feedback devices, and wearable technologies.

5.3 Liquid Crystal Elastomers (LCEs)

Liquid crystal elastomers are a class of smart materials that combine the properties of liquid crystals with elastomeric polymers. These materials exhibit unique deformation behaviour in response to external stimuli. LCEs possess a liquid crystal phase

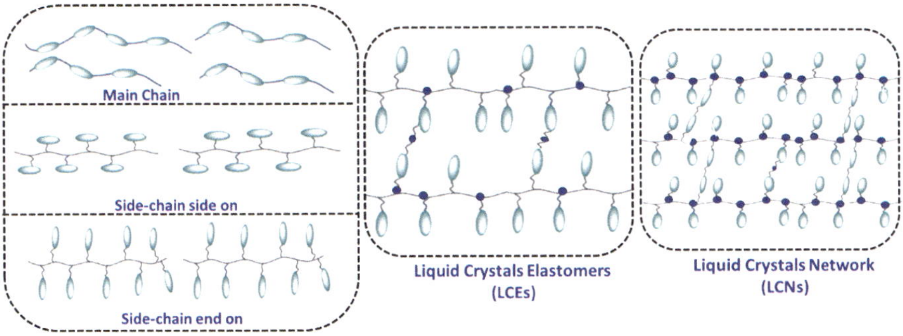

Fig. 4 Schematics showing a basic chain architectures of LCEs: main-chain, side-chain side-on/end-on, lightly crosslinked LCEs and densely crosslinked LCNs

that can change orientation in response to stimuli, resulting in macroscopic shape changes. For instance, photo-responsive LCEs can undergo reversible shape transformations when exposed to light. An example is the development of soft robotic actuators that leverage the shape-changing properties of LCEs to perform complex movements with high precision (Fig. 4). Responsive textiles made from LCEs can adjust their properties in response to environmental conditions, providing dynamic control of comfort and functionality in clothing. LCE-based artificial muscles can mimic the contraction and relaxation of natural muscles, offering potential applications in biomedical devices and prosthetics [72]. Additionally, LCEs are utilized in applications requiring programmable shape changes and high precision, such as soft robotics, adaptive optics, and responsive textiles.

5.4 Composite Smart Materials

Composite smart materials are integrating different types of materials to achieve enhanced properties and multifunctionality. These materials often combine polymers with inorganic fillers, such as metals or ceramics, to create composites that exhibit improved mechanical strength, electrical conductivity, or thermal stability [60]. The *shape memory polymer composites* combine a shape memory polymer matrix with reinforcing fibers or nanoparticles [54]. These composites can recover their original shape upon exposure to heat and also benefit from enhanced mechanical properties due to the reinforcing fillers. For example, the development of carbon fiber-reinforced polymer (CFRP) composites are used in the aerospace industry. These composites combine the high strength and stiffness of carbon fibers with the flexibility and lightweight nature of polymers, resulting in materials that are exceptionally strong yet lightweight. This combination significantly improves fuel efficiency and performance in aircraft. Additionally, in the medical field, biocompatible polymer composites reinforced with hydroxyapatite nanoparticles are used for bone implants.

These composites not only support the growth of natural bone tissue but also provide the necessary mechanical strength for the implants to function effectively. Another example is the use of graphene-reinforced polymers in flexible electronics, where the enhanced electrical conductivity and mechanical durability of graphene enable the creation of more robust and reliable electronic devices. Composite smart materials are employed in a wide range of applications, including aerospace, automotive, and medical devices, where both the smart functionality and the enhanced mechanical properties are crucial.

6 Smart Soft Materials and Their Functional Properties

Smart soft materials represent a cutting-edge class of materials that exhibit significant and reversible changes in their properties when exposed to external stimuli [2]. The dynamic response of these materials enables their use in a wide range of advanced applications, making them highly versatile and valuable materials for various industrial applications. The classification of smart soft materials based on four key functional properties: self-healing, shape-memory, conductive, and stimuli-responsive [3]. Each category highlights the unique mechanisms, applications, and examples of materials that fall under these classifications.

6.1 Self-healing Materials

Self-healing materials are engineered to repair damage caused by factors such as mechanical stress or aging. The healing process aims to restore the materials' properties, including mechanical strength, conductivity, protection, and aesthetics. That concept is not confined to a single class of materials but it is applicable across various types, including polymers and their composites, concretes and asphalts, as well as metals and ceramics. The development of self-healing polymers and composites often draws inspiration from natural materials, which inherently possess the ability to manage and repair damage throughout their lifetime [73].

Self-healing materials employ various mechanisms to repair damage, broadly categorized into intrinsic and extrinsic self-healing. Intrinsic self-healing involves the use of reversible interactions such as hydrogen bonding, ionic interactions, or dynamic covalent bonds that can break and reform, enabling the material to heal itself. For example, polyurethane-based coatings leverage dynamic covalent bonds to achieve intrinsic self-healing, which is beneficial in automotive and industrial applications where coatings frequently encounter scratches and minor damages [29]. Similarly, Diels–Alder polymers utilize reversible Diels–Alder reactions to facilitate self-healing, making them suitable for electronic devices and structural materials where maintaining integrity is crucial. On the other hand, extrinsic self-healing relies on embedded healing agents, such as microcapsules or vascular networks, which release a healing agent upon damage. These agents then repair the crack or damage. A notable example is microcapsule-embedded composites, where microcapsules containing healing agents are embedded in polymer matrices. When a crack propagates, the capsules rupture, releasing the healing agent to fill the crack and solidified by the catalyst embeded there and restore the material's integrity. This approach is widely used in construction of aerospace components to repair micro-cracks that can retain structural integrity. Both intrinsic and extrinsic self-healing mechanisms are pivotal in extending the lifespan and reliability of materials across various applications.

Self-healing materials are extensively used in coatings and paints in industries such as automotive and aerospace, where they help prolong the service life of components by autonomously repairing scratches and minor damages. Additionally, these materials are employed in biomedical implants to enhance their durability and reliability, thereby reducing the need for frequent replacements or repairs. The ability of self-healing materials to restore their original properties makes them invaluable in maintaining the functionality and longevity of various products and structures.

6.2 Shape-Memory Materials

Shape-memory materials (SMMs) possess the unique ability to recover their original shape from a deformed state when exposed to a specific stimulus, such as temperature or light. This remarkable property makes them highly valuable in applications requiring controlled and reversible shape change [44]. The concept of shape-memory effect is rooted in the material's ability to undergo a transformation between two distinct phases: a deformed phase and a recovered phase. The transition between these phases is triggered by external stimuli, leading to the material's ability to "remember" and revert to its original shape (Fig. 5). These are in two types mainly, first one is Shape-Memory Polymers (SMPs). These polymers can change shape in response to thermal stimuli. Upon heating above a certain temperature (transition temperature), they are returned to their pre-deformed shape. Another one is Shape-Memory Alloys (SMAs), typically made of nickel-titanium (NiTi), often called nitinol and these alloys respond to thermal stimulus [74]. They revert to their original shape upon heating due to phase transformation between austenite and martensite.

Mechanisms of shape-memory materials (SMMs) include thermo-responsive and photo-responsive mechanisms. In thermo-responsive SMMs, the material is deformed at a temperature below its transition temperature [59]. When the temperature is raised above this transition point, the material undergoes a phase change, returning to its original shape. This is the most widely used mechanism in both shape-memory alloys (SMAs) and shape-memory polymers (SMPs). For instance, NiTi alloys, which are widely used in medical stents and orthodontic wires due to their reliable shape-memory effect and biocompatibility, utilize this mechanism. Similarly, polyurethane as SMPs, used in self-deployable structures and textiles, leverage their low transition temperature and flexibility to achieve shape recovery

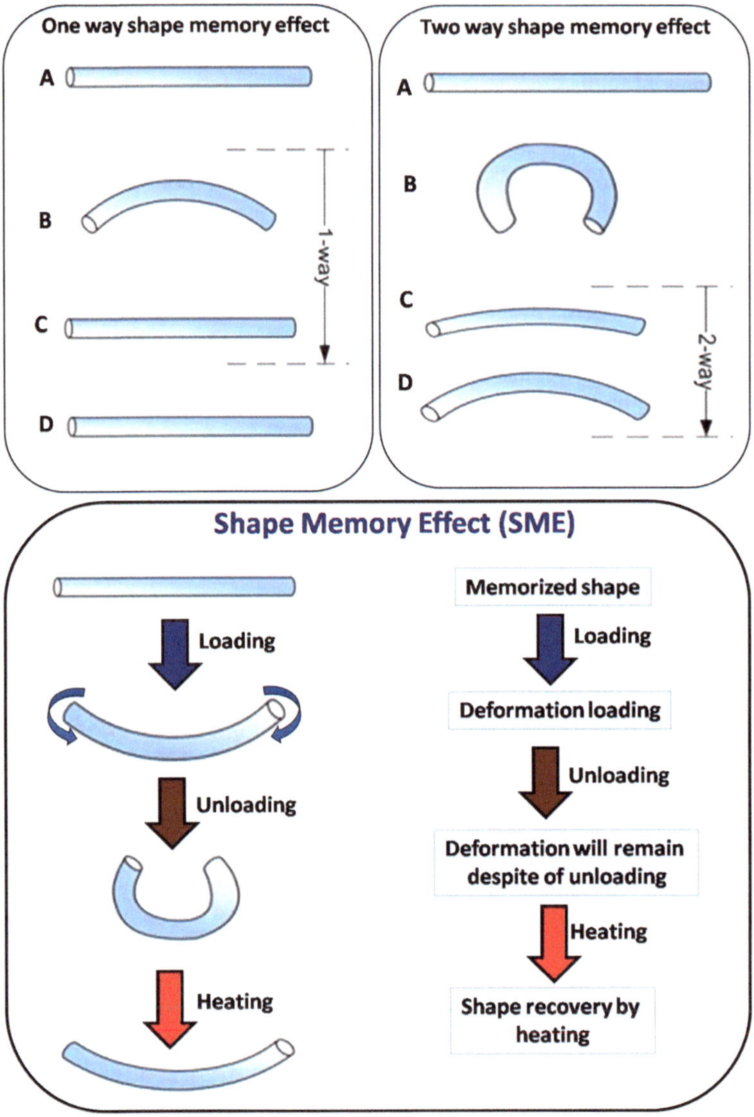

Fig. 5 Schematics showing the one-way SME and two-way SME and in the last showing loading and heating effect on SME

[29]. On the other hand, photo-responsive SMMs change shape in response to light exposure. These materials contain photo-responsive groups, such as azobenzene, into the polymer network, allowing them to undergo shape change when exposed to UV or visible light. Azobenzene-containing polymers are employed in light-responsive

actuators and soft robotics for their efficient photo-responsive shape change, enabling precise control over the shape transformation.

Shape-memory materials have diverse applications across various fields due to their unique ability to recover their original shape in response to specific stimuli. In medical devices, these materials adapt to body temperature, improving their functionality and reducing the need for frequent replacements. In textiles, self-adjusting clothing that responds to environmental temperature change enhances comfort and functionality. Aerospace and automotive industries utilize shape-memory materials in deployable structures and active suspension systems, enabling components to adjust based on operational conditions. Additionally, consumer electronics are benefited from shape-memory polymers in self-healing phone screens that maintain aesthetic quality and durability [68].

6.3 Conductive Materials

Conductive materials are essential for applications in electronics and electrochemical systems due to their ability to conduct electricity. These materials integrate the flexibility of polymers with the conductivity of metals or carbon-based materials, providing versatile solutions for various technological needs. Conductive polymers, such as polyaniline (PANI), are widely used in sensors, antistatic coatings, and as components in organic electronics due to their tunable electrical properties and ease of processing [50]. Graphene, known for its exceptional electrical conductivity and mechanical strength, is incorporated into polymer matrices to create nanocomposites used in flexible electronics, advanced batteries, and high-performance supercapacitors. PEDOT is another notable example, prized for its combination of high electrical conductivity and transparency, making it ideal for use in organic solar cells, touch screens, and flexible displays. The integration of these materials into electronic devices, energy storage systems, and smart textiles highlights their critical role in advancing technology and enhancing performance across a range of applications.

Conductive materials are of two types, (i) Conductive polymers are distinguished by their ability to conduct electricity due to their conjugated structures, which allow for the delocalization of electrons along the polymer backbone. Examples include polyaniline (PANI), poly(pyrrole) [50], and PEDOT. Polyaniline is widely used in sensors and antistatic coatings because of its tunable conductivity and ease of processing [49]. PEDOT is another significant conductive polymer, valued for its combination of high electrical conductivity and transparency, making it ideal for applications in organic solar cells and touch screens [45]. (ii) Nanocomposites materials enhance the electrical conductivity of polymers by incorporating conductive fillers such as carbon nanotubes, graphene, or metal nanoparticles. Graphene-polymer nanocomposites, for example, are employed in flexible electronics and high-performance batteries due to their superior conductivity and mechanical strength. By integrating these advanced fillers, nanocomposites achieve enhanced electrical and thermal properties, making them suitable for demanding technological applications.

Conductive materials play a crucial role in flexible electronics and energy storage systems, where their unique properties are essential for performance and functionality. In flexible electronics, materials like graphene-polymer nanocomposites are used in wearable devices and flexible displays, providing the necessary conductivity while allowing for bending and stretching. In energy storage, conductive polymers such as PEDOT are integrated into improved batteries and supercapacitors, enhancing their performance and efficiency due to their high conductivity and stability. These materials enable the development of advanced technologies that are both flexible and efficient, driving innovation in consumer electronics and energy solutions.

6.4 Stimuli-Responsive Materials

Stimuli-responsive materials exhibit changes in their physical or chemical properties in response to external stimuli [16]. These materials are designed to adapt their behavior dynamically based on environmental conditions, making them invaluable in various applications including sensors, actuators, and adaptive systems. Stimuli-responsive materials encompass a range of types that respond to different external stimuli, each with unique applications. Thermo-responsive polymers, such as poly(NIPAM), undergo a reversible transition from hydrophilic to hydrophobic state with temperature change, making them ideal for drug delivery systems and tissue engineering where temperature-induced swelling is beneficial [43]. pH-responsive polymers like poly(AA) adjust their solubility or volume in response to pH variation, which is useful in controlled drug release and bio separation processes. Photo-responsive polymers, such as those containing azobenzene, change their shape or state under UV or visible light, enabling applications in optical data storage and light-driven actuators. Electro-responsive polymers like polypyrrole modify their shape or conductivity when exposed to electric field, which is valuable in sensors and actuators. Magneto-responsive polymers, including magnetorheological elastomers, alter their stiffness in response to magnetic field, making them suitable for vibration damping and adaptive suspension systems in vehicles [69].

These are pivotal in drug delivery systems, enabling controlled release of drugs in response to specific physiological conditions, which enhances targeted therapy and minimizes side effects. In soft robotics, these materials are used in actuators and sensors that adapt to environmental stimuli, leading to more versatile and functional robotic systems [45].

7 Smart Soft Materials and Their Applications

Smart soft materials, characterized by their ability to respond dynamically to external stimuli, find diverse applications across several fields due to their unique functional properties. This classification highlights their roles in various domains.

7.1 Biomedical Applications

In the biomedical field, smart soft materials are revolutionizing drug delivery systems, tissue engineering, and medical implants. Thermo-responsive hydrogels, such as poly(N-isopropylacrylamide) (PNIPAM), control drug release based on body temperature, while pH-sensitive polymers like poly(acrylic acid) (polyAA) enable targeted drug delivery and improved patient outcomes [16]. Shape-memory polymers, such as those derived from polyurethanes and polycaprolactone (PCL), facilitate minimally invasive implantation and effective deployment by changing their shape at body temperature [29, 52]. Self-healing materials, including those made from polysaccharides like alginate combined with dynamic covalent bonds, repair damage autonomously, enhancing implant durability. Electro-responsive polymers, such as poly(pyrrole) and PEDOT, alter their properties in response to electrical stimuli, making them suitable for bioelectronic applications and neural interfaces. Additionally, dendritic polymers such as dendrimers offer high surface functionality for targeted drug delivery, and block copolymer micelles are used in controlled drug release and imaging applications due to their ability to encapsulate therapeutic agents [50].

7.2 Soft Robotics

Soft robotics are benefited significantly from smart materials that enable the creation of flexible, adaptive, and responsive robotic systems. Electro-responsive polymers, such as polyaniline (PANI) and conductive hydrogels based on chitosan, are utilized in actuators and sensors to achieve precise movements and interactions with their environment. Shape-memory alloys (SMAs), like Nickel-Titanium (NiTi), are employed to create adaptable robotic components that change shape in response to temperature change, mimicking natural movements [74]. Additionally, thermo-responsive hydrogels such as those based on poly(ethylene glycol) (PEG) are used to create soft actuators that expand or contract in response to temperature variation [34]. Magneto-responsive materials, including those with embedded magnetic nanoparticles, are applied in soft robots for applications requiring controllable actuation via external magnetic field [45]. Self-healing polymers, such as those made from dynamic covalent chemistry, enhance the durability and lifespan of soft robotic components by autonomously repairing damage. These smart materials enable soft robots to perform complex tasks in diverse environments, from delicated medical procedures to flexible manufacturing processes and dynamic exploration missions.

7.3 Wearable Technology

In wearable technology, smart soft materials are crucial in developing devices that offer both functionality and comfort. Conductive polymers, such as polyaniline (PANI) and polypyrrole, are widely used in smartwatches and fitness trackers to enable electronic components while maintaining flexibility and durability [20]. Piezoelectric materials, like zinc oxide (ZnO) nanowires and polyvinylidene fluoride (PVDF), are integrated into health-monitoring devices to convert mechanical stress into electrical signals, facilitating the accurate measurement of physiological parameters such as heart rate and body temperature [75]. Shape-memory polymers and thermo-responsive hydrogels are also employed to enhance the comfort and adaptability of wearable devices, allowing them to conform to the body's movements and adjust to temperature changes [34]. These smart materials ensure that wearables are not only functional but also provide real-time health insights, improving user experience and enabling better personal health management.

7.4 Adaptive Optics

These systems leverage smart materials to dynamically adjust optical properties in response to environmental conditions, enhancing system performance and precision. Photo-responsive polymers, such as those incorporating azobenzene or spiropyran, are used to create materials that change their optical properties, such as refractive index or color, when exposed to specific wavelengths of light [36]. Liquid crystal materials, like nematic and smectic phases, are employed in adaptive lenses and displays to provide real-time control over light transmission and optical alignment [37]. Additionally, electro-optic polymers such as polymer-dispersed liquid crystals (PDLCs) and poly(vinylidene fluoride) (PVDF) exhibit changes in optical properties under the influence of an electric field, making them suitable for high-precision optical applications. Phase-change materials, including vanadium dioxide (VO_2), which can switch between insulating and metallic states, are also utilized in adaptive optics for temperature-controlled optical switching and modulation. These materials enable high-precision control in various optical systems, including telescopes, high-resolution imaging devices, and adaptive lenses, improving their performance and adaptability in diverse conditions.

7.5 Smart Textiles

Smart textiles integrate responsive materials to create fabrics that interact dynamically with their environment, enhancing both functionality and aesthetics.

Thermo-responsive materials, such as phase-change materials (PCMs) and poly(N-isopropylacrylamide) (PNIPAM), are used in textiles to adjust insulation based on body temperature, providing adaptive comfort in varying thermal conditions [33]. Photochromic polymers, like azobenzene and spiropyran, enable fabrics to change color in response to light exposure, offering both functional and aesthetic versatility. Piezoelectric fibers, such as those made from polyvinylidene fluoride (PVDF), generate electrical signals when mechanically deformed, making them suitable for wearable sensors and energy-harvesting applications [27]. Conductive polymers, including polyaniline (PANI) and polypyrrole, are integrated into textiles to create smart clothing with embedded sensors for monitoring physiological parameters, such as heart rate and body temperature [49, 50]. Shape-memory alloys and thermo-responsive hydrogels are also utilized in textiles for their ability to alter their properties in response to temperature changes, enhancing the adaptability of the fabric for different applications [35]. These advanced materials contribute to the development of smart textiles that offer improved comfort, functionality, and adaptability in sportswear, fashion, and protective gear.

8 Conclusions

Smart Soft Materials (SSMs) are leading the way in material science and technology, representing a significant leap forward. This chapter provides a comprehensive overview of SSMs, organized into several key areas: responsive stimuli, composition, biomimetic materials, structural classification, applications, and functional properties. The chapter begins by detailing the different responsive stimuli that drive the behavior of SSMs, including thermo-responsive, photo-responsive, pH-responsive, magneto-responsive, electro-responsive, moisture-responsive, and mechanically responsive materials. It then categorizes SSMs based on their composition into polymeric smart materials (such as conductive, electroactive, and shape-memory polymers), hybrid smart materials (composite, multifunctional, and hybrid nanomaterials), and biomimetic smart materials inspired by natural systems.

Further, the chapter explores the structural classification of SSMs, highlighting hydrogels, elastomers, liquid crystal elastomers (LCEs), and composite smart materials. It examines their diverse applications in biomedical fields, soft robotics, wearable technology, adaptive optics, and smart textiles. The discussion includes specific examples of how these materials are applied in real-world scenarios, showcasing their versatility and impact. The chapter concludes with an analysis of the functional properties of SSMs, such as self-healing, shape-memory, conductive, and stimuli-responsive behaviors, underscoring their potential to revolutionize various technological domains. By integrating theoretical insights with practical examples, this chapter provides a holistic view of SSMs and their transformative role in modern technology.

References

1. Miyata, T.: Polym. J. **42**, 277–289 (2010)
2. Ramakrishnan, T., Kumar, S.S., Chelladurai, S.J.S., Gnanasekaran, S., Sivananthan, S., Geetha, N.K., Arthanari, Assefa, B.G.: J. Nanomater. **25**, 9185–9198 (2022)
3. Ghosh, A., Nag, S., Gomes, A., Gosavi, A., Ghule, G., Kundu, A., Purohit, B., Srivastava, R.: Micromachines **14**, 121–145 (2023)
4. Coyle, S., Majidi, C., LeDuc, P., Hsia, K.J.: Extreme Mech. Lett. **22**, 51–59 (2016)
5. El-Atab, N., Mishra, R.B., Al-Modaf, F., Joharji, L., Alsharif, A.A., Alamoudi, H., Diaz, M., Qaiser, N., Hussain, M.M.: Adv. Intell. Syst. **2**, 2000128 (2020)
6. Yu, C.Y., Mu, J.H., Fu, Y.L., Zhang, Y.C., Han, J.S., Zhao, R.Y., Zhao, J., Wang, Z.H., Zhao, Z.C., Li, W.J., Liu, F.S.: Chin. J. Polym. Sci. **39**, 417–424 (2021)
7. Wang, B., Guo, Z., Liu, W.: RSC Adv. **4**, 14684 (2014)
8. Lucarini, S., Hossain, M., Garcia-Gonzalez, D.: Compos. Struct. **279**, 114800 (2022)
9. Singh, V., Prasad, Y.S., Rachamalla, A.K., Rebaka, V.P., Banoo, T., Maheswari, C.U., Sridharan, V., Lalitha, K., Nagarajan, S.: RSC Adv. **12**, 2022 (2009)
10. Dixit, A., Bag, D.S.: Polymer **119**, 263–273 (2022)
11. Schwartz, M., Lagerwall, J.P.F.: Build. Environ. **226**, 109714 (2022)
12. Bocchetta, G., Fiori, G., Sciuto, S.A., Scorza, A.: Actuators **12**, 261 (2023)
13. Ramakrishnan, T., Kumar, S.S., Chelladurai, S.J.S., Gnanasekaran, S., Sivananthan, S., Geetha, N.K., Arthanari, R., Assefa, G.B.: J. Nanomater. **14** (2022). https://doi.org/10.1155/2022/4031059
14. Chen, S., Wang, H.-Z., Liu, T.-Y., Liu, J.: Adv. Intell. Syst. **5**, 2200375 (2023)
15. Bahl, S., Nagar, H., Singh, I., Sehgal, S.: Mater. Today Proc. **28**, 1302–1306 (2020)
16. Ghosh, A., Nag, S., Gomes, A., Gosavi, A., Ghule, G., Kundu, A., Purohit, B., Srivastava, R.: Micromachines **14**, 121 (2023)
17. Zhao, X., Chen, X., Yuk, H., Lin, S., Liu, X., Parada, G.: Chem. Rev. **121**, 4309–4372 (2021)
18. Hu, Z B., Cai, T., Chi, C L., Soft Matter **6**, 2115-2123 (2010)
19. Fuentes, E., Gerth, M., Berrocal, J.A., Matera, C., Gorostiza, P., Voets, I.K., Pujals, S., Albertazzi, L.: J. Am. Chem. Soc. **142**, 10069–10078 (2020)
20. Rad, J.K., Balzade, Z., Mahdavian, A.: J. Photochem. Photobiol. C Photochem. Rev. **51**, 100487 (2022)
21. Dixit, A., Bag, D.S., Sharma, D.K., Prasad, N.E.: Polym. Int. **68**, 503–515 (2019)
22. Carpa, R., Farkas, A., Dobrota, C., Butiuc-Keul, A.: Gels **9**, 278–289 (2023)
23. Dudchenko, N., Pawar, S., Perelshtein, I., Fixler, D.: Materials **15**, 2601 (2022)
24. Lee, L., Yoo, D., Ling, D., Cho, M.H., Hyeon, T., Cheon, J.: Chem. Rev. **115**, 10637–10689 (2015)
25. Joaquim, A., Paul, O., Ibezim, M., Johnson, D., Falconer, A., Wu, Y., Williams, F., Mu, R.: Polymers **13**, 2702 (2022)
26. Liang, Y., Qiao, L., Qiao, B., Guo, B.: Chem. Sci. **14**, 3091–3116 (2023)
27. Chang, C., Zhang, L.: Carbohydr. Polym. **84**, 40–53 (2011)
28. Wang, M., Bai, J., Shao, K., Tang, W., Zhao, X., Lin, D., Huang, S., Chen, C., Ding, Z., Ye, J.: Int. J. Polym. Sci. **42**, 1–16 (2021)
29. Gupta, A., Maharjan, A., Kim, B.S.: Appl. Sci. **21**, 4694 (2019)
30. Golinelli, N., Spaggiari, A., Dragoni, E.: J. Intell. Mater. Syst. Struct. **28**, 953–960 (2017)
31. Zhou, C., Wu, T., Xie, X., Song, G., Ma, X., Mu, Q., Huang, Z., Liu, X., Sun, C., Xu, W.: Eur. Polym. J. **177**, 111454 (2022)
32. Qiao, S., Wang, H.: Nano Res. **10**, 5400–5423 (2018)
33. Wu, S., Hu, W., Ze, Q., Sitti, M., Zhao, R.: Multifunct. Mater. **3**, 042003 (2020)
34. Ward, M.A., Georgiou, T.K.: Polymers **3**, 1215–1242 (2011)
35. Meng, Q., Hu, J.: Sol. Energy Mater. Sol. Cells **92**, 1260–1268 (2008)
36. Ye, N., Pei, Y., Han, Q., Jin, L.Y.: Soft Matter **19**, 1540 (2023)
37. Matsuda, K., Irie, M.: J. Photochem. Photobiol. C **5**, 169–182 (2004)

38. Rose, M.J., Mascharak, P.K.: Coord. Chem. Rev. **252**(18–20), 2093–2114 (2008)
39. Unyong, J., Yin, Y.: Adv. Funct. Mater. **30**, 1907059 (2020)
40. Bonferoni, M.C., Caramella, C., Catenacci, L., Conti, B., Dorati, R., Ferrari, F., Genta, I., Modena, T., Perteghella, S., Rossi, S., Sandri, G., Sorrenti, M., Torre, M.L., Tripodo, G.: Pharmaceutics **13**, 1341 (2021)
41. El-Husseiny, H.M., Mady, E.A., Doghish, A.S., Zewail, M.B., Abdelfatah, A.M., Noshy, M., Mohammed, O.A., El-Dakroury, W.A.: Int. J. Biol. Macromol. **260**, 129323 (2024)
42. Chakroborty, S., Nath, N., Mahal, A., Barik, A., Panda, A.R., Fahaduddin, F., Bal, T., Obaidullah, A., Elawady, A.J.: J. Mol. Liq. **403**, 124828 (2024)
43. Fragal, E.H., Fragal, V.H., Silva, E.P., Paulino, A.T., da Silva Filho, E.C., Mauricio, M.R., Silva, R., Rubira, A.F., Muniz, E.C.: Carbohydr. Polym. **292**, 119665 (2022)
44. Wang, H., Yang, L., Yang, Y., Zhang, D., Tian, A.: Chem. Eng. J. **469**, 143976 (2023)
45. Xu, Y., Jia, Y., Liu, P., Jiang, Q., Hu, D., Ma, Y.: Chem. Eng. J. **404**, 126552 (2021)
46. Youn, J.H., Jeong, S.M., Hwang, G., Kim, H., Hyeon, K., Park, J., Kyung, K.U.: Appl. Sci. **2**, 640 (2020)
47. Chowdhury, S., Mishra, A., Venkateswaran, P., Somvanshi, D., Mater. Sci. Semicond. Process. **163**, 107572 (2023)
48. Mishra, A., Kuanr, B.K., Mohanty, T.: AIP Conf. Proc. **1832**, 1 (2017)
49. Lo, M., Ktari, N., Gningue-Sall, D., Madani, A., Aaron Efremova, S., Aaron, J.-J., Mekhalif, Z., Delhalle, J., Chehimi, M.: Emerg. Mater. **3**, 815–839 (2020)
50. Baker, C.O., Huang, X., Nelson, W., Kaner, R.B.: Chem. Soc. Rev. **46**, 1510–1525 (2017)
51. Wang, W., Lv, Y., Liu, H., Cao, Z.: Sep. Purif. Technol. **330**, 125265 (2024)
52. Saxman, A.M., Liepins, R., Aldissi, M.: Prog. Polym. Sci. **11**, 57–89 (1985)
53. Muscas, F., Sessini, V., Peponi, L., López, A.J., Ureña, A., Navarro, R., Marcos-Fernández, Á.: Polymers **17**, 3447 (2022)
54. Basak, S., Bandyopadhyay, A.: Polym. Adv. Technol. **33**, 2091–2112 (2022)
55. Mishra, A., Mohanty, T.: AIP Conf. Proc. **1731**, 1 (2016)
56. Luo, L., Zhang, F., Leng, J.: Compos. Sci. Technol. **213**, 108899 (2021)
57. Özkan, D., Gok, M.S., Karaoglanli, A.C.: Adv. Compos. Mater. **12**, 385–405 (2020). https://doi.org/10.1007/978-3-030-39062-4_20
58. Morampudi, P., Namala, K.K., Gajjela, Y.K., Barath, M., Prudhvi, G.: Mater. Today Proc. **43**, 314–319 (2021)
59. Venditti, I.: Polymers **14**, 2022 (1901)
60. Zang, G., Zhou, L., Zhang, X., Pan, D., Wang, M., Luo, S., Su, F., Ji, Y., Liu, C.: Compos. Commun. **37**, 101450 (2023)
61. Zhi, M., Liu, Q., Chen, H., Chen, X., Feng, S., He, Y.: ACS Omega **4**, 10975–10984 (2019)
62. Mishra, A., Yadav, N., Bhattacharya, J., Moulick, R.G.: J. Appl. Phys. **125**, 16 (2019)
63. Mushahary, N., Sarkar, A., Basumatary, F., Brahma, S., Das, B., Basumatary, S.: Results Surf. Interfaces **15**, 100225 (2024)
64. Guo, F., Aryana, S., Han, Y.: Appl. Sci. **8**, 1696 (2018)
65. Goetjen, T.A., Liu, J., Wu, Y., Sui, J., Zhang, X., Hupp, J.T., Farha, O.K.: Chem. Commun. **56**, 10409–10418 (2020)
66. Li, J., Li, S., Huang, J., Khan, A.Q., An, B., Zhou, X., Liu, Z., Zhu, M.: Adv. Sci. **9**, 2103965 (2022)
67. Seidi, K., Ayoubi-Joshaghani, M.H., Azizi, M., Javaheri, T., Jaymand, M., Alizadeh, E., Webster, T.J., Yazdi, A.A., Niazi, M., Hamblin, M.R., Amoozgar, Z., Jahanban-Esfahlan, R.: Nano Today **39**, 101157 (2021)
68. Vitalis, T., Gross, A., Gerasimidis, S.: J. Appl. Mech. **919**, 091007 (2024)
69. Bahl, S., Nagar, H., Singh, I., Sehgal, S.: Mater. Today Proc. **28**, 1302–1306 (2020)
70. Zhao, X., Ming, H., Wang, Y., Luo, F., Li, Z., Li, J., Tan, H., Fu, Q.: ACS Appl. Bio Mater. **4**, 5352–5361 (2021)
71. Samaha, M.A., Vahedi Tafreshi, H., Gad-el-Hak, M.: C. R. Mecanique **340**, 18–34 (2012)
72. Dixit, A., Bag, D.S.: Eur. Polym. J. **175**, 111347 (2022)
73. Carpi, F., De Rossi, D.: Mater. Sci. Eng. C **24**, 555–562 (2004)

74. Kularatne, R.S., Kim, H., Boothby, J.M., Ware, T.H.: J. Polym. Sci. Part B Polym. Phys. **55**, 395–411 (2017)
75. Cordier, P., Tournilhac, F., Soulié-Ziakovic, C., Leibler, L.: Nature **451**, 21 (2008)

Liquid Crystal Spatial Light Modulators for Beam Shaping and Wavefront Sensing

Nagendra Kumar, Santanu Konwar, S. S. Goutam Buddha, Inderpreet Kaur, and Karuna S. Malik

Abstract Liquid Crystal Spatial Light Modulators (LCSLM) are devices capable of spatially and temporally modulating the amplitude and phase of incident light beams, offering versatile applications in optics. This chapter focuses on the primary applications of LCSLMs, such as beam manipulation using Computer-Generated Holography (CGH) techniques. The dynamic modulation capabilities of LCSLMs enable the implementation of CGH for wavefront sensing, which detects and measures beam aberrations. We introduce the use of LCSLMs for measuring and correcting aberrations in light beams, thereby enhancing the wavefront quality. Additionally, we present simulation and experimental results that highlight the performance of LCSLMs in these applications. Furthermore, we discuss the integration of a wavefront sensor using LCSLM, which allows for real-time measurement of wavefront aberrations.

Keywords Liquid crystal · Polarization · Phase · Light waves

1 Introduction

Liquid crystal spatial light modulators (LCSLMs) have become indispensable in modern optics, offering unparalleled control over the fundamental properties of light waves. These devices dynamically modulate light in amplitude, phase, or polarization, enabling various scientific and technological applications. Their ability to

N. Kumar (✉) · I. Kaur
Department of Physics, Indian Institute of Technology Delhi, New Delhi, India
e-mail: nagendra123iitg@gmail.com

S. Konwar
Department of Physics, Abhayapuri College, Abhayapuri, Bongaigaon, Assam, India

S. S. Goutam Buddha
Department of Physics, Tata Institute of Fundamental Research, Hyderabad, India

K. S. Malik
Department of Physics, Brainware University, Barasat, Kolkata, West Bengal, India

A. Mishra and V. Pathak (eds.), *Smart Materials for Energy Storage and Biomedical Applications*, Engineering Materials, https://doi.org/10.1007/978-3-031-92584-9_8

manipulate light with high precision has revolutionized fields ranging from adaptive optics and holography to advanced imaging, quantum computing, and optical communication systems [1–6].

In amplitude modulation, LCSLMs adjust the light intensity profile by selectively transmitting or attenuating portions of the incident beam. This fine-grained intensity control is critical for binary and grayscale pattern generation applications, where pixel-level precision ensures optimal system performance [7–10]. Optical pattern generation and structured light applications depend heavily on this functionality to create high-resolution spatial intensity distributions. Amplitude modulators are also extensively used in display technologies, lithography, and data encoding, where precise illumination of spatial domains is fundamental to high contrast and accuracy.

In phase modulation, LCSLMs alter the wavefront of light, providing control over the optical path length and phase. This capability is central to holography, where LCSLMs generate interference patterns for creating intricate and dynamic three-dimensional images. Phase modulation is equally crucial in adaptive optics systems, which rely on real-time wavefront correction to improve image quality and enhance the performance of telescopes, microscopes, and other high-resolution imaging devices. Additionally, phase modulation plays a pivotal role in beam steering, enabling precise redirection of light beams in lidar systems, free-space optical communication, and optical trapping applications [11–15]. This functionality is also critical in quantum computing, where LCSLMs facilitate precise control of atomic arrays [16, 17] and in advanced microscopy techniques, which aid in high-resolution biological imaging [18].

Polarization modulation, another key capability of LCSLMs, enables dynamic adjustment of the polarization state of light. This feature is especially valuable in advanced imaging systems and polarization-sensitive communication technologies. For instance, differential interference contrast (DIC) microscopy leverages polarization modulation to enhance the visualization of refractive index variations within samples, offering superior image clarity. In optical communication, polarization modulation encodes and decodes information, improves data integrity, and optimizes system performance. One of the most distinctive advantages of LCSLMs is their ability to operate in both static and dynamic modes, enabling real-time control of light's properties. This versatility is particularly significant for adaptive optical systems, where dynamic modulation is essential for wavefront sensing and aberration correction applications [19, 20]. For example, LCSLMs facilitate thin-film thickness characterization and wavefront analysis in precision metrology, which demands extreme accuracy and reliability [19]. Furthermore, LCSLMs have found emerging applications in fields such as quantum technologies, which are used for controlling and manipulating quantum states, and in autonomous systems, where they enhance lidar and imaging capabilities. The adaptability and multifunctionality of LCSLMs underscore their transformative impact across diverse optical science and technology domains.

This chapter aims to comprehensively explore the principles and applications of LCSLMs, emphasizing their critical role in beam manipulation, adaptive optics, advanced imaging, and precision wavefront measurements.

2 Working Principle of LCSLM

Liquid crystal spatial light modulators (LCSLMs) operate on the principles of liquid crystal technology to achieve precise light modulation. These devices exploit the unique dual-phase properties of liquid crystals, which exhibit characteristics of both liquids and solid crystals. The liquid crystal molecules are highly sensitive to external electric fields. In their default state (without an applied electric field), liquid crystal molecules are aligned in a specific orientation, determined by boundary conditions at the surfaces of the cell. When an electric field is applied, the orientation of the liquid crystal molecules changes, thereby altering the optical properties of the light interacting with them. This phenomenon enables the precise manipulation of light, facilitating applications such as beam shaping, wavefront control, and real-time adaptive optics.

The performance of an LCSLM is predominantly determined by two critical factors: resolution and speed. The resolution is dictated by the number of pixels in the liquid crystal array, with higher resolutions allowing for finer control over the spatial modulation of light. This is essential for applications like high-resolution holography and advanced imaging systems. Speed, or the response time of the liquid crystal molecules to changes in the electric field, is influenced by the material properties of the liquid crystals and the driving electronics. Faster response times are indispensable for applications requiring real-time modulation, such as adaptive optics and high-speed optical communication.

Liquid crystal molecules are of three major types: nematic, smectic, and cholesteric. The Nematic Liquid Crystals (NLCs) are the most common type, where molecules are aligned in random directions but tend to be parallel. The orientation of the molecules can be changed by applying an electric field across the liquid crystal volume. In the smectic phase, the molecules are arranged in layers, with all the molecules in each layer having a specific orientation. In the cholesteric phase, the liquid crystal molecules are again arranged in the layers, showing that the orientation of the molecules changes from layer to layer in a helical fashion. Figure 1(i)–(iii) show the typical arrangements of molecules in the nematic, smectic and cholesteric phases, respectively.

Moreover, LCSLMs are used in various optical applications, such as optical communication, where light signals are dynamically modulated and enable high-speed data transmission in holography to generate and reconstruct holograms by modulating the phase or amplitude of the light waves, allowing for real-time holographic displays; in adaptive optics systems to correct wavefront distortions in real-time, improving the performance of telescopes and other imaging systems; in microscopy, for advanced imaging techniques such as confocal microscopy and structured illumination microscopy, enhancing resolution and contrast. Here, we present the fundamental application of LCSLM, focusing on holographic display. Here, we explore the mechanisms, technical specifications, and specific applications that play a pivotal role in wavefront shaping and sensing. Figure 2 illustrates the principle behind beam shaping using a pixelated device. Where it shows that the entire area of

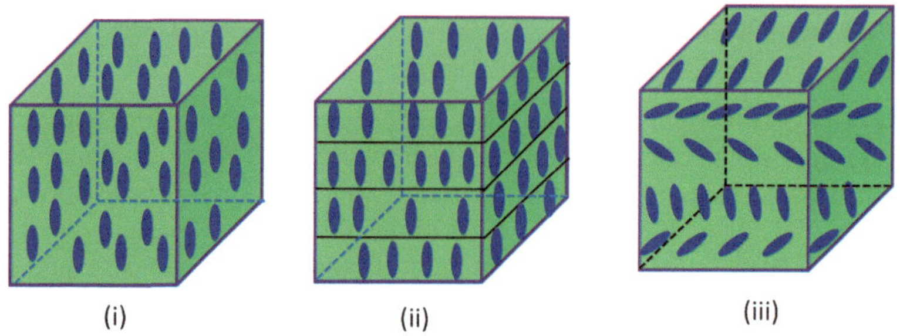

Fig. 1 Arrangements of the LC cells in the case of **(i)** nematic liquid crystal, **(ii)** smectic liquid crystal and **(iii)** cholesteric liquid crystal

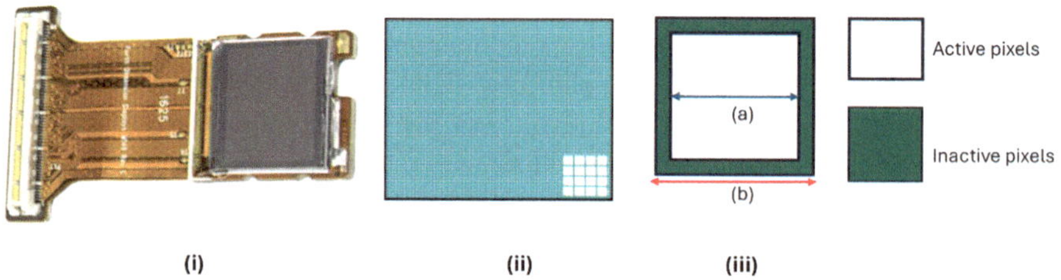

Fig. 2 **(i)** Snapshot of a pixel from a spatial light modulating device; **(ii)** numerically simulated array of pixels, with an inset showing a magnified view of the pixel arrangement; **(iii)** (**a**) represents the width of the optically active part, (**b**) denotes the width of the optically inactive part, with the magnified view adjacent

the pixel of a pixelated device is not optically active, as each pixel must accommodate supporting electronics. The white colour shows the active pixel, and the green colour represents the inactive pixels.

3 Interaction of Gaussian Beam with SLM and Resulting Diffraction Pattern

In this experiment, a Gaussian beam is initially directed onto a spatial light modulator (SLM), where its spatial characteristics are modulated according to a predefined pattern. The SLM alters the phase and amplitude of the incident beam, causing it to diffract. Following modulation, the diffracted beam passes through a lens. The lens focuses on the diffracted light, transforming the spatially varying intensity distribution into a series of focal spots at its focal plane. This focusing effect results on a diffraction pattern that reflects the SLM's modulation profile, revealing the interaction between the incident beam and the optical elements. The pattern and distribution of the focal spots provide insights into the modulation's efficiency and

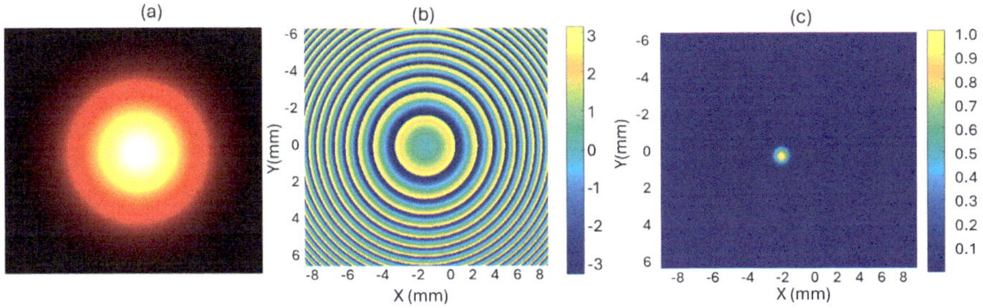

Fig. 3 **a** Intensity profile of the Gaussian beam incident on the spatial light modulator (SLM); **b** phase pattern of the beam after modulation by the SLM, followed by the diffraction of the beam as it passes through a lens; **c** focal spots formed by the lens focusing the diffracted beam

the lens system's optical performance. We have observed that as we increase the initial beam diameter of the incident beam size of the focal spot pattern decreases. This can be understood through diffraction theory. The PSF of an optical system is largely determined by the diffraction of light as it passes through the aperture. The diffraction angle is inversely proportional to the aperture size. The diffraction angle θ is inversely proportional to the aperture diameter DD: θ ~ λ/D where the λ wavelength is. A larger aperture reduces θ, resulting in a less spread-out beam, but broadens the PSF due to wavefront confinement. Therefore, a larger beam diameter leads to a broader PSF, which means the imaging system has a lower resolving power (Fig. 3).

4 Generating a Vortex Beam Using a Phase Spatial Light Modulator

A vortex beam, characterized by its helical wavefront and orbital angular momentum, can be generated by encoding a specific phase profile onto a spatial light modulator (SLM). The key to creating a vortex beam lies in imparting a phase singularity, which introduces a phase twist across the beam's profile. This phase profile is designed to induce a helical phase structure in the transmitted or reflected light, resulting in a beam with a well-defined vortex structure. The phase profile $\phi(r, \theta)$ applied to the SLM must introduce a phase discontinuity that varies azimuthally around the beam's center to create a vortex beam. The phase profile for the *lth* order vortex beam is given by: $\phi(r, \theta) = l\theta$ where, r is the radial distance from the beam's center, θ is the azimuthal angle, and l is the topological charge or vortex order, determining the number of helical twists.

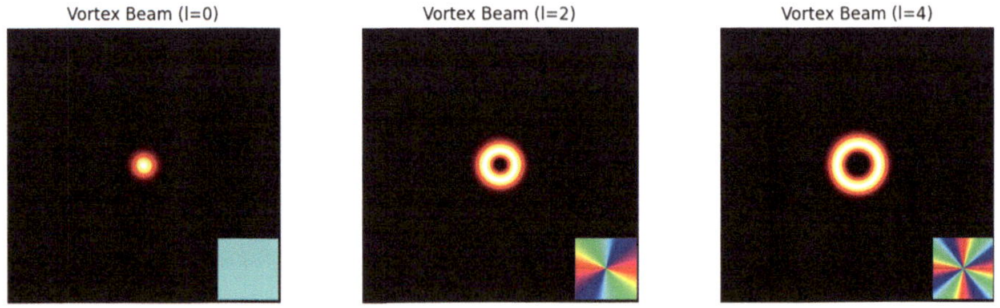

Fig. 4 Intensity and corresponding phase distribution of vortex beams with topological charge of $l = 0, 2, 4$

4.1 Encoding the Phase Profile on the SLM

The SLM modulates the phase of the incident Gaussian beam by applying the phase profile $\phi(r, \theta)$. The spatial light modulator can be programmed to impose this phase shift across the beam's profile. If the SLM has a phase modulation range of 2π the phase profile can be encoded using:

$$\phi(r, \theta) = \frac{2\pi l\theta}{\lambda}$$

where the wavelength of the incident beam is, this phase modulation introduces a helical phase front in the outgoing beam, resulting in the vortex beam. The optical field $E(r, \theta)$ of the vortex beam, after passing through the phase-modulated SLM, can be expressed as:

$$E(r, \theta) = A(r)e^{i\phi(r,\theta)}e^{ikz}$$

where A(r) is the radial amplitude profile of the beam (often a Gaussian profile), k is the wave number, and z is the propagation distance along the beam axis. The resulting vortex beam will exhibit a phase singularity at the center, where the intensity pattern typically features a central dark spot surrounded by a ring of higher intensity. The beam's helical phase front creates an angular momentum that the vortex charge (l) quantifies (Fig. 4).

5 Numerical Approach to Generate a Vortex Beam

This section outlines key techniques for generating vortex beams, including optical elements like spiral phase plates, computer-generated holography (CGH), digital devices such as spatial light modulators, and digital micromirror devices. Figure 5 shows a computer-generated binary hologram and corresponding diffraction orders.

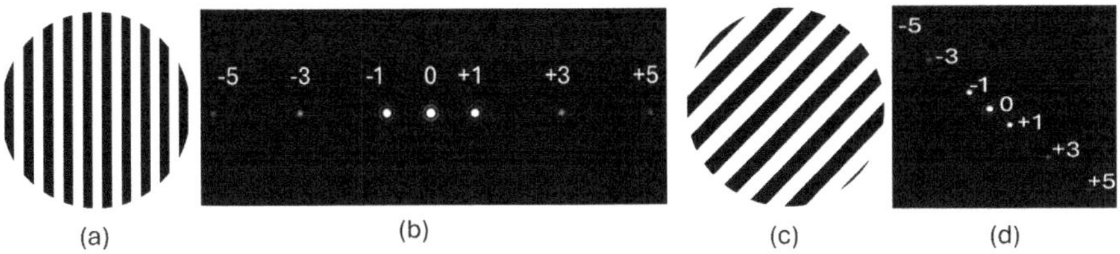

Fig. 5 Images of amplitude holograms using **a** mx = 10π, my = 0 and **c** mx = 5π, my = 5π. **b**, **d** The intensity distribution in the Fourier plane of the binary holograms (**a**) and (**c**), respectively

Now, we focus on generating a vortex beam by integrating CGH with SLM. More details about the CGH can be obtained in [21–23]. Holographic optical elements such as fork grating and spiral zone plates are commonly used to generate optical vortices [24, 25]. To obtain the desired intensity profile light, the interference pattern of a plane wave with a complex amplitude of

$$\Phi_1(x, y) - A_1(x, y)e^{i\phi(x,y)}$$

and the vortex light with complex amplitude of

$$\Phi_2(x, y) = A_2(x, y)e^{il\theta}$$

Taking the superposition of two waves and interference intensity is

$$I = E_1(x, y) + E_2(x, y)e^{i[\phi(x,y)-l\theta]} + E_2(x, y)e^{-i[\phi(x,y)-l\theta]}$$

where $E_1(x, y) = |A_1(x, y)|^2 + |A_2(x, y)|^2$ and $E_2(x, y) = |A_1(x, y)||A_2(x, y)|$. Then, the lithography technique binarises the fork-shaped grating and records it in a photographic film. Suppose Φ_1(x, y) as the reference wave for example, a Gaussian beam, then the following beams can be generated: $E_1(x, y)e^{i\phi(x,y)} + E_2(x, y)e^{i[2\phi(x,y)-l\theta]} + E_2(x, y)e^{l\theta}$ where the first term represents an unmodulated plane wave, the second term is the combination of vortex and reference wave and the third term is the desired vortex wave. Figure 6 shows the numerically simulated fork gratings and their PSF in the Fourier plane having the topological charges of 1 and 2. We can see the size of the doughnut increases as we increase the topological charges. The size of the doughnut in a vortex beam increases with higher topological charges because the beam's phase twist becomes more pronounced. Higher topological charges (l) impart greater angular momentum, causing the light's intensity to spread radially. This results in a larger doughnut radius, scaling approximately as $R \propto \sqrt{1}$, due to the increased spatial distribution of phase and energy.

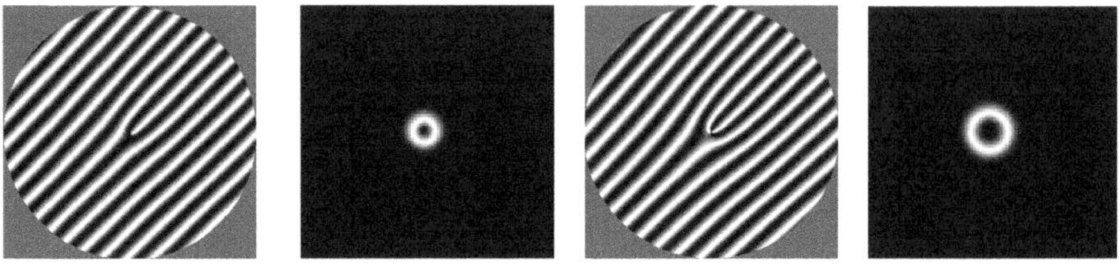

Fig. 6 Holograms were designed for orbital angular momentum (OAM) vortex wave generation, which had topological charges of 1 and 2 and corresponding point spread functions in the Fourier plane

6 Experimental Arrangements to Generate User-Defined Wavefront Utilizing the SLM

In practice, the vortex beam is generated by projecting the modulated phase profile onto the SLM, which then modulates the incident Gaussian beam. After modulation, the beam passes through a lens to focus on the beam for specific applications. By carefully designing the phase profile and encoding it onto the SLM, it is possible to generate vortex beams with different orders and characteristics, allowing for various optical manipulation, imaging, and communication applications. Figure 7 shows that an incident beam is directed towards a spatial light modulator (SLM), which modulates the light to produce various diffraction orders. The diffracted light is then collected by a lens (L1), which focuses it onto an iris diaphragm. The diaphragm is strategically positioned to block all diffraction orders except the + 1-diffraction order. This selected + 1 order is subsequently passed through and focused by a second lens (L2) and finally, imaged onto a detector (P1P2), allowing for the analysis of the diffraction pattern and the spatial information encoded by the SLM.

7 Beam Correction Using Spatial Light Modulators (SLMs)

The advent of spatial light modulators (SLMs) and computer-generated holography (CGH) has ushered in a new era of precision in beam correction and wavefront shaping. These technologies enable dynamic control over laser beams' amplitude, phase, and polarization, providing a powerful platform for optimizing beam quality, compensating aberrations, and generating complex optical patterns tailored to specific applications.

Utilizing the property of SLM and vortex point spread function, we have used it as a tool for beam correction, particularly in real-time optical systems. When a Gaussian beam passes through an optical system, any distortions—such as defocus, astigmatism or optical misalignment occur onto the SLM where fork grating is displayed,

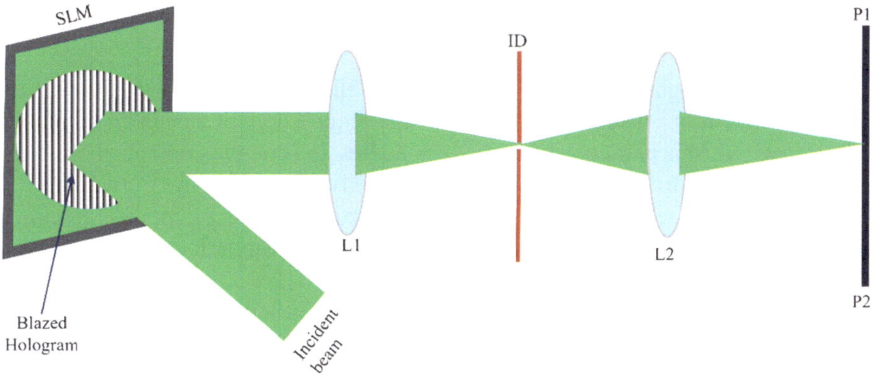

Fig. 7 Basic optical setup for diffracted beam analysis: an incident beam impinges on a spatial light modulator (SLM), which diffracts the light. A lens (L1) collects the diffracted beams, while an iris diaphragm blocks all diffraction orders except the + 1 order. These + 1 orders is then focused by a second lens (L2) and imaged onto the detector (P1P2)

and it causes the vortex core to shift or deform. Analyzing these changes in the vortex structure makes it possible to identify the type and extent of the aberrations. In the experiment, our aim is to correct the beam by creating the perfect doughnut structure, providing a reliable means for real-time monitoring and correction of the optical system. Figure 8a illustrates a PSF of Gaussian beam incident onto the SLM, which has an astigmatic aberration, and we tried to make it count structure to correct them as illustrated in Fig. 8b. After the beam correction, we can see the aberrated free PSF of the Gaussian beam into the detector in the Fourier plane. Moreover, adaptive optics systems use this information to dynamically adjust the wavefront, compensating for aberrations and restoring the ideal PSF. In addition to wavefront correction, vortex PSFs are particularly useful for improving imaging quality and precision in applications such as microscopy, optical trapping, and laser machining.

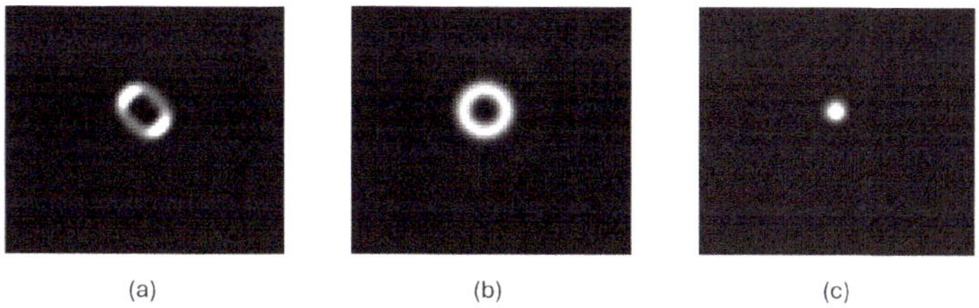

Fig. 8 Beam correction using vortex beam as a reference tool

8 Wavefront Sensing with Spatial Light Modulators

This section highlights the flexibility and adaptability of SLM, making it ideal for real-time wavefront sensing in applications that require high precision and dynamic adjustments. SLM is particularly effective at addressing complex optical aberrations in systems ranging from high-power lasers to adaptive optics. It plays a crucial role in wavefront sensing, where they correct beam distortions, and in biomedical imaging, where they improve clarity by compensating for optical imperfections. A detailed description of wavefront sensing can be found in Refs. [26–30].

To validate the wavefront correction and sensing scheme, a proof-of-principle experiment is conducted, as shown in the schematic diagram (Fig. 9a). A He–Ne red laser is used as the illumination source, with lenses L1 and L2 expanding and collimating the beam before it strikes the SLM. The SLM modulates the phase of the incident light, displaying either an array of binary diffraction gratings or a single hologram. Lens L3 collects the diffracted light, and an iris diaphragm isolates the + 1-diffraction order. Lens L4 focuses the light onto a CMOS camera. The experiment demonstrates the system's ability to manipulate light in a controlled manner, with diffraction patterns analyzed to validate its functionality. Real-time beam correction is shown in the left column of Fig. 9b, with the point spread function displayed before and after correction. The right column of Fig. 9b shows the applied and estimated phase profiles.

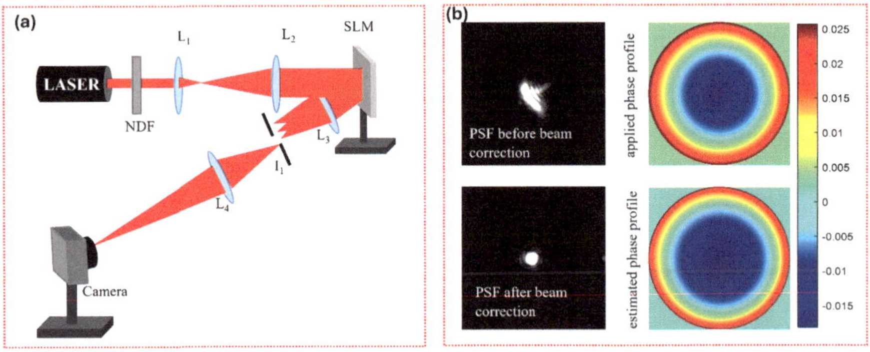

Fig. 9 **a** Schematic of wavefront sensing using SLM, **b** real-time beam correction and wavefront measurement

9 Conclusion

In this chapter, we explored the working principles of liquid crystal spatial light modulators (LCSLMs), highlighting their ability to dynamically modulate light in amplitude, phase, and polarization. Their unique properties and versatile functionality make them indispensable for various optical applications, including beam shaping, beam correction, and wavefront sensing. We validated the theoretical principles through experimental setups, demonstrating the practical implementation of LCSLMs in beam correction and wavefront measurement. The experimental results confirmed the capability of LCSLMs to correct beam aberrations and precisely measure wavefront distortions, emphasizing their significance in adaptive optics and advanced imaging systems. Overall, this chapter illustrates the transformative potential of LCSLMs in optical technology, providing a foundation for their continued development and application in cutting-edge scientific and industrial domains.

References

1. Bruggeman, D.A.G.: Optical modulators for advanced optical communication systems. J. Lightwave Technol. **32**(6), 1141–1153 (2014)
2. Robinson, M.D., Lane, J.R., Watkins, K.G.: Adaptive optics using LCSLMs: a review. Opt. Express **22**(11), 13498–13512 (2017)
3. Vecchi, L.M.: Advances in liquid crystal technology for photonics. J. Appl. Phys. **111**(3) (2019)
4. Kagalwala, K.H., et al.: Single-photon three-qubit quantum logic using spatial light modulators. Nat. Commun. **8**(1), 739 (2017)
5. Konwar, S., Boruah, B.R.: Leveraging the orthogonality of Zernike modes for robust free-space optical communication. Commun. Phys. **3**(1), 203 (2020)
6. Kumar, N., et al.: Computer generated holography: techniques, applications, and future directions. In: Materials for Electronic, Magnetic, and Spintronic Technologies: Characterization and Applications from Energy Storage to Disease Detection, pp. 153–169. Springer Nature Switzerland, Cham (2024)
7. Kumar, N., Khare, A., Boruah, B.R.: Enhanced dynamic range of the grating array based zonal wavefront sensor using a zone wise scanning method. In: Photonic Instrumentation Engineering VII, vol. 11287. SPIE (2020)
8. Kumar, N., et al.: Improvement in wavefront measurement accuracy in grating array based zonal wavefront sensors by modulating the intensity profile of the incident beam. In: Proceedings of SPIE, vol. 12445
9. Goutam Buddha, S.S., Boruah, B.R.: Estimation of point spread function of an optical microscope using stochastic minimization of least square errors. J. Opt. **22**(5), 055603 (2020)
10. Kalita, R., Goutam Buddha, S.S., Boruah, B.R.: Suitability of holographic beam scanning in high resolution applications. In: Three-Dimensional and Multidimensional Microscopy: Image Acquisition and Processing XXV, vol. 10499, p. 104991P (2018)
11. Malik, K.S., Boruah, B.R.: Optimal trap velocity in a dynamic holographic optical trap using a nematic liquid crystal spatial light modulator. J. Opt. **24**(3), 034004 (2022)
12. Sherpa, N., et al.: Synchronization and clock recovery in a ferroelectric liquid crystal spatial light modulator based free-space optical communication link. Rev. Sci. Instrum. **94**(5) (2023)
13. Malik, K.S., Kumar, N., Boruah, B.R.: Dynamic modulation of spatial intensity profile of a laser beam using a binary hologram. Opt. Commun. **515**, 128201 (2022)

14. Malik, K.S., et al.: Estimation of maximum translation speed of a trapped particle in holographic optical tweezers. In: Complex Light and Optical Forces XIV, vol. 11297. SPIE (2020)
15. Malik, K.S., Boruah, B.R.: Optical force calculation in the ray-optics regime for beams with arbitrary complex amplitude profiles. Opt. Lett. **47**(16), 4151–4154 (2022)
16. Kim, H., et al.: Gerchberg-Saxton algorithm for fast and efficient atom rearrangement in optical tweezer traps. Opt. Express **27**(3), 2184–2196 (2019)
17. Yun, X., et al.: Zero-order free holographic optical tweezers. Opt. Express **31**(12), 19613–19621 (2023)
18. Nikolenko, V., et al.: SLM microscopy: scanless two-photon imaging and photostimulation using spatial light modulators. Front. Neural Circuits **2**, 393 (2008)
19. Kumar, N., Khare, A., Boruah, B.R.: Enhancement of wavefront measurement sensitivity in a zonal wavefront sensor without curtailing the sensing speed. J. Opt. **24**(1), 014003 (2021)
20. Kumar, N., Khare, A., Boruah, B.R.: Enhanced dynamic range of the grating array based zonal wavefront sensor using a zone wise scanning method. In: Photonic Instrumentation Engineering VII, vol. 11287. SPIE (2020)
21. Kumar, N., et al.: Experimental demonstration of in situ surface and thickness profile measurements of thin film during deposition using a grating array based wavefront sensor. Opt. Lett. **47**(21), 5509–5512 (2022)
22. Hariharan, P.: Basics of Holography. Cambridge University Press (2002)
23. Tricoles, G.: Computer generated holograms: an historical review. Appl. Opt. **26**(20), 4351–4360 (1987)
24. Cirino, G., et al.: Digital holography: computer-generated holograms and diffractive optics in scalar diffraction domain (2011)
25. Guo, C.-S., et al.: Optical vortex phase-shifting digital holography. Opt. Express **12**(21), 5166–5171 (2004)
26. Wang, D., et al.: Generating arbitrary arrays of circular Airy Gaussian vortex beams with a single digital hologram. Appl. Phys. B **127**, 1–5 (2021)
27. Kumar, N.: Development of grating array based zonal wavefront sensor for in situ surface profiling during the growth of the thin film in a deposition system (2022)
28. Platt, B.C., Shack, R.: History and principles of Shack-Hartmann wavefront sensing. J. Refract. Surg. **17**(5), S573–S577 (2001)
29. Zhao, L., et al.: Efficient implementation of a spatial light modulator as a diffractive optical microlens array in a digital Shack-Hartmann wavefront sensor. Appl. Opt. **45**(1), 90–94 (2006)
30. Zepp, A., et al.: Simulation-based design optimization of the holographic wavefront sensor in closed-loop adaptive optics. Light Adv. Manuf. **3**(3), 384–399 (2022)

Hydrostatic Pressure Effects on Transmittance and Reflectance Spectra in 1D Defective Semiconductor Photonic Crystals Under TE Mode Configuration

Chhaya Bhatia, Yogesh Sharma, Mirza Tanweer Ahmad Beig, Shikha Shukla, Rajiv Vatsa, Akanksha Namdeo, Sanjeev K. Srivastava, Amodini Mishra, and A. N. Upadhyay

Abstract In this study, we theoretically investigate the effect of hydrostatic pressure on the reflectance and transmittance properties of the one dimensional (1D) Defective Semiconductor Photonic Crystal (DSPhC) containing Germanium (Ge) and Silicon (Si). In this study the 1D Photonic Crystal (PhC) is composed of alternate layers of Ge and Si having a finite number of layers. In order to calculate the transmittance and reflectance spectra of the proposed PhC, the transfer matrix method [TMM] has been used. Using a theoretical model based on a TMM for both normal and oblique incidence angles transmission and reflectance spectra are obtained for the 1D DSPhC.

Keywords Hydrostatic pressure · Semiconductor photonic crystal · Photonic band gap

---

C. Bhatia · Y. Sharma (✉) · M. T. A. Beig · R. Vatsa · A. Namdeo
Department of Physics, Faculty of Applied and Basic Sciences, SGT University, Gurugram, Haryana, India
e-mail: uvsbhu@gmail.com

S. Shukla
Department of Physics, Institute of Sciences, Banaras Hindu University, Varanasi, U.P., India

S. K. Srivastava
Department of Physics, Amity Institute of Applied Sciences, Amity University Uttar Pradesh, Noida, India

A. Mishra
Department of Physics, Indian Institute of Technology Delhi, New Delhi, India

A. N. Upadhyay
Department of Physics, Lady Irwin College, University of Delhi, New Delhi, India

A. Mishra and V. Pathak (eds.), *Smart Materials for Energy Storage and Biomedical Applications*, Engineering Materials, https://doi.org/10.1007/978-3-031-92584-9_9

1 Introduction

Photonic crystals (PhCs) have emerged as a transformative class of optical materials due to their unique ability to control and manipulate the flow of light through periodic modulation of the refractive index. One-dimensional (1D) photonic crystals, in particular, offer a simpler geometry and ease of fabrication while retaining key photonic properties such as photonic band gaps (PBGs), which prohibit light propagation within specific frequency ranges [1]. Including defect layers within these periodic structures introduces localised states inside the bandgap, giving rise to defect modes that enable the design of narrowband optical filters, sensors, and resonators [2, 3]. Semiconductor materials are often employed in PhCs due to their high refractive indices, tunability, and compatibility with optoelectronic applications [4]. When combined with external perturbations such as hydrostatic pressure, these semiconductor photonic structures exhibit significant changes in their optical response. Hydrostatic pressure alters constituent materials' lattice parameters and refractive indices, thereby modifying the photonic band structure and defect mode characteristics [5, 6]. The resistance-strain effect of materials [7, 8], the piezoelectric effect [9–11], or mechanical elastic deformation [12, 13] are the basic operating principles of traditional pressure sensors. This tunability is particularly advantageous for pressure-sensing applications and the design of adaptive photonic devices. While much of the existing literature has focused on temperature and electric field effects on PhCs [14, 15], the role of hydrostatic pressure remains relatively underexplored, especially in defective 1D semiconductor photonic crystals under transverse electric (TE) mode illumination. TE mode analysis, in which the electric field is perpendicular to the direction of propagation, is crucial for understanding polarisation-dependent behaviour in optical devices [16]. Investigating how hydrostatic pressure influences transmittance and reflectance spectra under TE mode provides deeper insight into the photonic behaviour of these structures. It opens pathways for engineering pressure-tunable optical filters and sensors. This chapter presents a comprehensive study on the effects of hydrostatic pressure on the optical properties, specifically the transmittance and reflectance spectra, of 1D defective semiconductor photonic crystals. Using the transfer matrix method (TMM), we analyse how varying pressure impacts the position and intensity of defect modes under TE polarisation. The results reveal critical insights into the interaction between mechanical and optical phenomena in PhCs and underscore their potential in developing next-generation pressure-sensitive photonic devices.

The contents of this research chapter are prearranged as follows; Sect. 1 is Introduction, Sect. 2 is structural design, Theoretical model and approach. Section 3 is result and discussion. Finally, Sect. 4 is the conclusion.

2 Theoretical Model and Approach

The schematic diagram of 1D-DSPhC is given as Fig. 1. In this we consider the two layers named as A and B, where A represents the Germanium (Ge) and B represents the Silicon (Si) having refractive indices as n_A and n_B and thickness d_A and d_B. 1D-DSPhC under consideration is depicted as $air/(AB)^N A/air$ respectively. $d = d_A + d_B$ represents the period of structure and N number of periods of each structure. In order to calculate the reflectance and transmittance spectra of the proposed 1D-DSPhC, the TMM is used [17], which is very effective technique to study the transmission and reflectance properties of finite 1D-DSPhC.

Let an electromagnetic [EM] wave in the air be injected into a 1D-DSPhC at an incident angle in the + x direction. The electric and the magnetic field at the two interfaces of each layer are correlated by TMM. The characteristic matrix M_K for the Kth layer in the PhC for transverse electric waves can be given as [18].

$$M_K = \left(\cos\cos\delta_K \frac{1}{iq_K} \sin\sin\delta_K - iq_K \sin\sin\delta_K \cos\cos\delta_K \right) \tag{1}$$

where $\delta_K = \frac{2\pi}{\lambda} n_k d_k \cos\cos\theta_K$, $K = A, B$, θ_K the ray is angle inside the layer K of refractive index m and λ is the wavelength of light in the incidence medium and $\cos\cos\theta_K = \sqrt{1 - \left(n_o^2 \sin^2\theta_o / n_K^2\right)}$ where $q_K = n_K \cos\cos\theta_K$ for TE.

In this the numerical result of reflection and transmission property of the 1D-DSPhC taken in the form $air/(AB)^N A/air$ is presented. The total characteristics matrix for the proposed 1D-DSPhC can be given by N period of the system are given by:

$$M = (M_A M_B)^N M_A \tag{2}$$

where, M_A and M_B are the characteristics matrices of layer A and B respectively.

The reflection and transmission coefficient for TE waves is given as [12],

$$r = \frac{(M_{11} + q_t M_{12}) q_o - (M_{21} + q_t M_{22})}{(M_{11} + q_t M_{12}) q_o + (M_{21} + q_t M_{22})} \tag{3}$$

$$t = \frac{2q_o}{(M_{11} + q_t M_{12}) q_o + (M_{21} + q_t M_{22})} \tag{4}$$

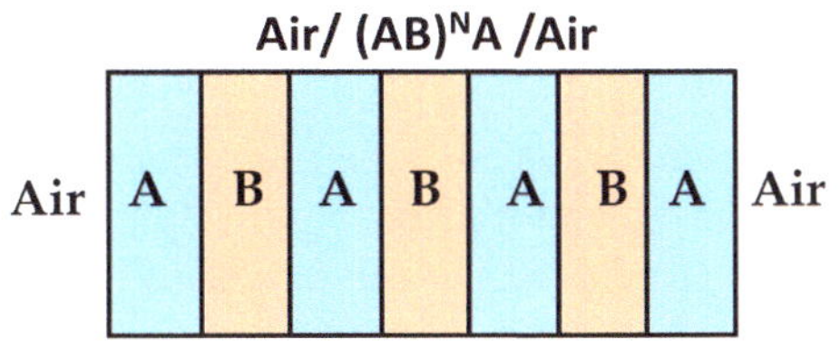

Fig. 1 Schematic diagram of the proposed 1D-DSPhC

where M_{11}, M_{12}, M_{21} and M_{22} are the elements of the total characteristic matrix of N period multilayer structures. The values of q_o and q_t for TE polarized wave is given as:

$$q_o = n_o \cos\cos\theta_o; \quad q_t = n_t \cos\cos\theta_t \tag{5}$$

Here n is the refractive index of the substrate and θ is the ray angle inside it. Finally, the reflectance and the transmittance of the proposed 1D-DSPhC structure can be obtained by using the expression:

$$R = |r|^2 \tag{6}$$

$$T = \frac{q_t}{q_o}|t|^2 \tag{7}$$

3 Result and Discussion

The hydrostatic pressure (P) dependent refractive index of Ge is given as [19]

$$n_A = \sqrt{15.94 - 0.36P + 0.014P^2} \tag{8}$$

Figure 2 illustrates the variation in germanium's refractive index under hydrostatic pressures between 0 and 5 Pa. The refractive index of Ge is calculated in accordance with Eq. 1 and refractive index of Si is 3.5. The calculated value of refractive index of Ge at different P ranges from 0 to 5 Pa is given in Table 1. It is observed that the refractive index of Ge decreases with the increase in pressure.

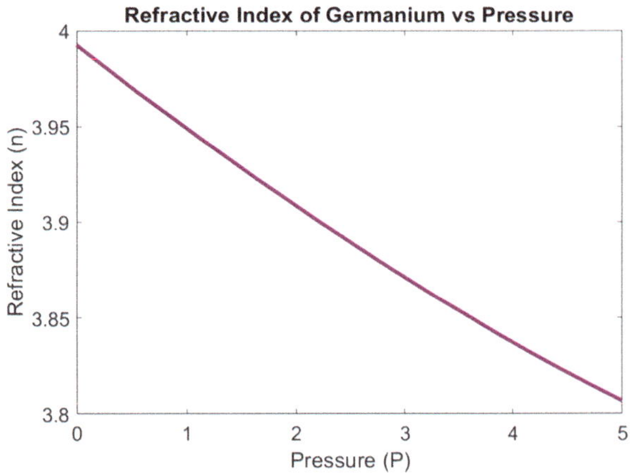

Fig. 2 Variation of the refractive index of Ge with applied hydrostatic pressure (P)

Table 1 Variation of refractive index of Ge with hydrostatic pressure

Hydrostatic pressure P (Pa)	0	1	2	3	4	5
Refractive index	3.992	3.949	3.908	3.871	3.837	3.807

Effect of Hydrostatic Pressure on Reflectance and Transmittance Properties of 1D-DSPhC

The hydrostatic pressure range of 0–5 Pa and the normal incidence angle for the 1D-DSPhC with $d_A = 97.1$ nm, $d_B = 94.4$ nm and N = 14 have plotted and are displayed below. The transmittance spectra for different values of applied pressure for different values of applied pressure P in the range 0–5 Pa have been plotted below. The thickness of the constituent layers are taken according to the quarter wave stack condition, $n_A d_A = n_B d_B = \frac{\lambda_o}{4}$ where $\lambda_o = 1500$ nm is the central wavelength, so the thickness of Ge layer $d_A = 97.1$ nm and d_B (Si) = 94.4 nm (Fig. 3).

The transmission peak is located at 1475, which is precisely identical to the central wavelength, at P = 0, according to the transmittance curve analysis. The transmission peak moves to the lower side of the wavelength as the hydrostatic pressure increases. The medium that allows light to travel through becomes denser as pressure rises. The transmission intensity may drop as a result of increased light scattering and absorption brought on by this greater density.

We may deduce from the analysis of these curves that when the hydrostatic pressure rises, the width of the reflection bands decreases. Ge layer's changing refractive index is the cause of the variation in the width of the reflection band. Because Ge exhibits a lower refractive index value at higher pressures, the hydrostatic pressure causes a reduction in the contrast between the refractive index values of the constituent layers. The breadth of the reflection bands decreases as the refractive index differences decrease. Additionally, it has been observed that the reflection band moves toward the wavelength's lower end as the applied pressure increases. Furthermore, at greater pressure levels, the variation in the breadth of the subsequent reflection bands is far less pronounced.

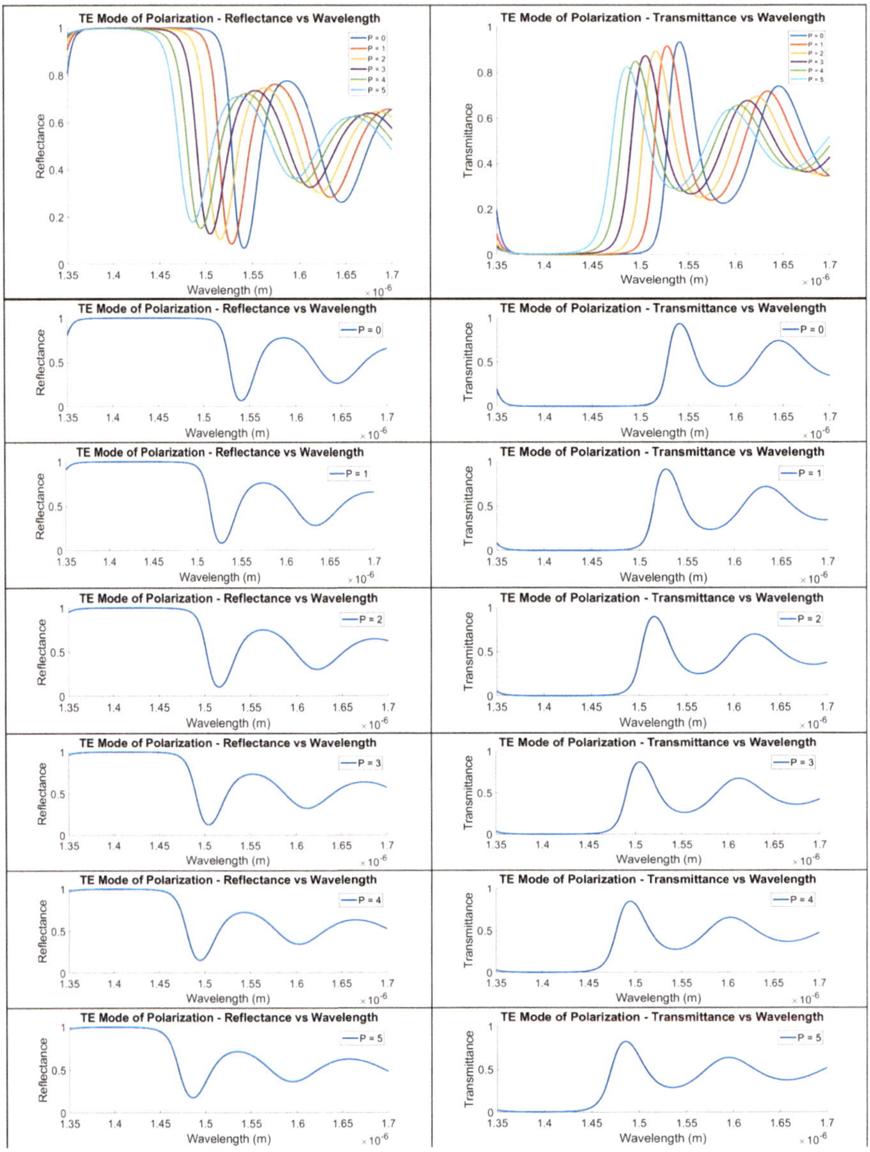

Fig. 3 Reflectance and transmission curves of 1D-DSPhC structure with d_A (Ge) = 97.1 nm and d_B (Si) = 94.4 nm for hydrostatic pressure ranging from 0 to 5 Pa

4 Conclusion

In conclusion, the analysis shows that the transmission peak aligns with the central wavelength at P = 0 but shifts toward lower wavelengths as hydrostatic pressure increases. This shift is caused by increased medium density, which results in more light scattering and absorption, thereby reducing transmission intensity. The reduction in the width of reflection bands under pressure is attributed to the decreasing refractive index contrast between layers, especially as Ge exhibits a lower refractive index at higher pressures. Consequently, the reflection bands also shift toward shorter wavelengths, with less variation in their widths at higher pressure levels.

References

1. Joannopoulos, J.D., Johnson, S.G., Winn, J.N., Meade, R.D.: Photonic Crystals: Moulding the Flow of Light, 2nd ed. Princeton University Press (2008). https://doi.org/10.2307/j.ctvcm4gz9
2. Yariv, A., Yeh, P.: Photonics: Optical Electronics in Modern Communications, 6th ed. Oxford University Press (2007). https://librarysearch.wlv.ac.uk/permalink/44UOWO_INST/8bamsd/alma991001691109704901
3. Sakoda, K.: Optical Properties of Photonic Crystals, 2nd ed. Springer (2005). https://doi.org/10.1007/b138376
4. Russell, P.S.J.: Photonic crystal fibres. Science. https://doi.org/10.1126/science.1079280
5. Villeneuve, P.R., Fan, S., Joannopoulos, J.D.: Microcavities in photonic crystals: Mode symmetry, tunability, and coupling efficiency. https://doi.org/10.1103/PhysRevB.54.7837
6. Mehaney, A., Ahmed, A.M., et al.: Hydrostatic pressure effects for controlling the phononic band gap properties in a perfect phononic crystal (2022). https://doi.org/10.1007/s11082-021-03484-9
7. Mohammed, A.A., Haris, S.M., et al.: Using the pressure transmission coefficient of a transmitted wave to evaluate some of the mechanical properties of refractory metals. Ultrasonics **55**, 133–140 (2015). https://doi.org/10.1016/j.ultras.2014.07.001
8. Kulkarni, A., Kim, H., et al.: A novel approach to use of elastomeric for monitoring of pressure using plastic optical fiber. Rev. Sci. Instrum. **81**, 01597 (2010). https://doi.org/10.1063/1.3386588
9. Roszkiewicz, A., Garlińska, M., Pregowska, A.: Advancements in Piezoelectric-Enabled Devices for Optical Communication (2024). https://doi.org/10.1002/pssa.202400298
10. Veeralingam, S., Bharti, D.K., et al.: Lead-free PDMS/PPy-based low-cost wearable piezoelectric nanogenerator for self-powered pulse pressure sensor application. Mater. Res. Bull. **151**, 111815 (2022). https://doi.org/10.1016/j.materresbull.2022.111815
11. Li, X., Wang, K., et al.: Plantar pressure measurement system based on piezoelectric sensor: a review. Sens. Rev. **42**, 241–249 (2022). https://doi.org/10.1108/SR-09-2021-0333
12. Paliwal, S., Yenuganti, S., et al.: Fabrication and testing of a Hall effect-based pressure sensor. Sens. Rev. **42**, 354–364 (2022). https://doi.org/10.1108/SR-01-2022-0001
13. Wang, H., Zou, D., et al.: A novel high-sensitivity MEMS pressure sensor for rock mass stress sensing. Sensors **22**, 7593 (2022). https://doi.org/10.3390/s22197593
14. Li, Z.Y., Lin, L.L., Zhang, Z.Q.: Spontaneous emission from photonic crystals: full vectorial calculations. Phys. Rev. Lett. **84**(21), 4341–4344. (2000). https://doi.org/10.1103/PhysRevLett.84.4341
15. Sherawat, V., R. Bokolia, Sinha, R.K.: Studies on temperature-dependent bandgap and gap-to-midgap ratio in diamond lattice photonic crystal with biosensing applications (2023). https://doi.org/10.1016/j.optmat.2023.114470

16. Gulia, A., Beig, M.T., Sharma, B., Sharma, Y.: Study of radiation power spectra in the 1-D photonic periodic dielectric structure (2023). https://doi.org/10.1088/1742-6596/2426/1/012023
17. Segovia-Chaves, F., Vinck-Posada, H.: Tuning of transmittance spectrum in a one-dimensional superconductor-semiconductor photonic crystal. Phys. B Condens. Matter **543**, 7–13 (2018). https://doi.org/10.1016/j.physb.2018.05.005
18. Goi, A.R., Syassen, K., et al.: Effect of pressure on the refractive index of Ge and GaAs. Phys. Rev. B **41**, 10104–10110 (1990). https://doi.org/10.1103/PhysRevB.41.10104
19. Born, M., Wolf, E.: Principles of Optics. Cambridge University Press, Cambridge (1999). https://doi.org/10.1017/CBO9781139644181

Understanding Photoluminescence Spectroscopy: Principles, Applications, and Insights into Material Properties

Shivam Swaraj and Amodini Mishra

Abstract This chapter gives an overview of photoluminescence, explaining the basic principles by its theories and through several approaches, such as steady-state and time-resolved photoluminescence, which depict the latest improvements and enhancements in the sensitivity, resolution, and fields of application of PL measurements. PL applies to a very large variety of materials: semiconductors, nanomaterials, organic molecules, and hybrid perovskites. The dynamics of excitations, as well as defect states, and the bandgap energies, are very important parameters of semiconductor research that need to be derived for the maximum efficiencies of electrical and optoelectronic devices. Information about size-dependent properties and knowledge about surface states on nanomaterials like carbon nanotubes, quantum dots, and nanowires are provided by PL, and it offers unprecedented opportunities at the frontier of science for the development of nanoscale sensors, light-emitting devices, flat panel displays, and biomedical imaging tools in nanotechnology. The utilization of PL enables the pharmaceutical industry to characterize different aspects of drug delivery systems, the interactions of biomolecules, and biosensor development. For instance, organic photovoltaics (OPVs) and organic light-emitting diodes (OLEDs) in organic electronics can optimize the luminous properties of organic materials with the help of PL analysis. The environmental sector takes advantage of PL in pollution identification and measurement, water quality assessment, and the development of novel materials for environmental remediation. Thus, photoluminescence provides an extremely powerful and versatile tool that is important for most fields of science. In such a context, this chapter seeks to underline the critical role of PL in pushing the boundaries of scientific research and industrial applications by focusing on the major ways of its usage and demonstrating the impacts on material characterization, product development, and technological innovation.

S. Swaraj
Central Research Facility, Sonipat Campus, Indian Institute of Technology Delhi, New Delhi, India

A. Mishra (✉)
Department of Physics, Indian Institute of Technology Delhi, Hauz Khash, New Delhi, India
e-mail: amodiniphysics@gmail.com

A. Mishra and V. Pathak (eds.), *Smart Materials for Energy Storage and Biomedical Applications*, Engineering Materials, https://doi.org/10.1007/978-3-031-92584-9_10

Keywords Photoluminescence · Nanotechnology · Hybrid perovskites · Organic photovoltaics (OPVs) and organic light-emitting diodes (OLEDs)

1 Introduction

Photoluminescence, often referred to as photoexcitation, is a non-destructive analytical technique that induces photon absorption, a radiative process when the material is illuminated with an ionization source mainly a laser or xenon lamp resulting in luminescence. Luminescence in light-emitting material results from a radiative process called photon emission, and there are several electronic excitation processes involved, including excitonic transition, intrinsic electronic transition between energy bands, and extrinsic electronic transition showing impurities and defects in organic molecules, semiconductors, and insulators. While measuring a PL in sophisticated instruments like steady state photoluminescence or time resolve photoluminescence spectrophotometer, a plot will be recorded, i.e., emitted light intensity versus wavelength of ionization source, which is discussed further in this chapter.

Photoluminescence for the first time was "officially" observed and inscribed in 1603 by Vincenzo Cascariolo from the outskirts of Bologna, Italy. The effect came to be discovered, surprisingly, through the use of a naturally occurring stone called a baryte or barite that, when heated and then exposed to sunlight, luminesced. This stone is nowadays called the Bologna Stone or Bolognian Stone [1]. In the mid-nineteenth century, Fluorescence was termed by G. G. Stokes (created Stokes law: scattered light always has a longer wavelength than incident light), and he observed scattered light in multiple directions [2]. Henri Becquerel (who discovered the photovoltaic effect) the 1st one who observed that the emitted light had a longer wavelength than the incident radiation. When calcium sulfide is deposited on paper that has already been exposed to solar light, emission of light occurs beyond the violet part of the spectrum, and this phenomenon was described by Becquerel in one of his papers [3]. So, Stoke's experiment was relevant to fluorescence and Becquerel's experiment was relevant to phosphorescence, and both are co-related to the PL.

Photoluminescence (PL) is classified into fluorescence, phosphorescence, and delayed fluorescence, distinguished by the duration of light emission after excitation. Fluorescence ceases immediately when excitation ends, while phosphorescence continues emitting light after excitation stops. This definition is not proper according to the criteria because there is the existence of short-lived phosphorescence, like fused quartz, and long-lived fluorescence, such as metal–ligand emitter and many quantum dots with hundreds of nanoseconds comparable lifetimes. In 1929 Francis Perrin observed phosphorescence and reported that before emission [4], the excited-state species transitions through an intermediate state. In molecular photochemistry, fluorescence retains spin multiplicity, while phosphorescence involves a transition between a triplet state and a singlet state (Fig. 1). Among the applications of photoluminescence that had an important impact in the field of spectroscopy are fluorescence microscope (Time-resolved fluorescence microscope), fluorescent lighting,

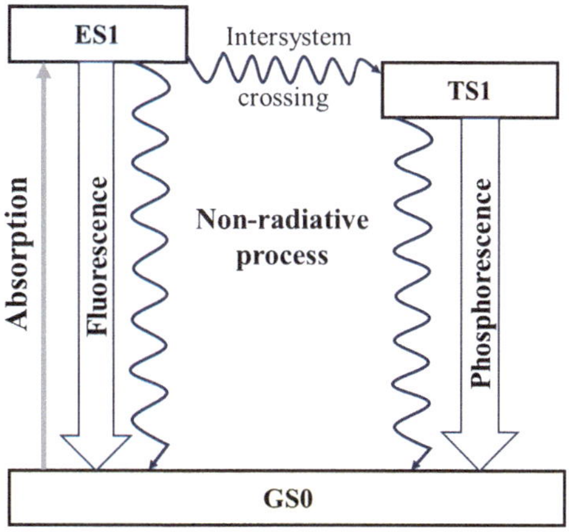

Fig. 1 The distinction between fluorescence and phosphorescence is emphasized in the Perrin–Jablonski diagram. Radiation is released from photon excitation at singlet state (ES1), attained from photon absorption resulting fluorescence, and radiation is released from the triplet state (TS1), via intersystem crossing from ES1, resulting phosphorescence

Fig. 2 Portrait of Sir George Gabriel Stokes (left) and Edmond Becquerel (right)

optical brighteners, plasma displays, forensic analysis, hydrogeological tracers, fluorescent and phosphorescent paints, safety signage, and the detection of counterfeits of security documents, banknotes, and works of art [5, 6].

2 Fundamentals Principles of Photoluminescence

Photoluminescence (PL) occurs when a material absorbs photons from an excitation source. The absorbed photon transfers its energy to an electron, promoting it from the ground state to an excited state within a femtosecond timescale—a process governed by quantum mechanics. In organic molecules, processes like thermalization or internal conversion occur between states with the same spin multiplicity. These processes take place when multiple excited states are available, allowing the excited electron to relax non-radiatively to lower excited states. This relaxation happens through molecular vibrations in organic molecules or the emission of phonons (discrete quasiparticles) in solids and occurs within picoseconds. The electron eventually recombines with the ground state, emitting photons in the process. The emitted photon has lower energy compared to the absorbed photon, which characterizes the photoluminescence phenomenon [7]. The energy difference between orbitals or bands can be determined by measuring the energy of the emitted photons, which provides information about the HOMO–LUMO gap in molecules or the direct band gap in semiconductors. In solids, the valence band (VB) edge and conduction band (CB) edge are analogous to the HOMO and LUMO states, respectively. As a result, semiconductors and ionic crystals exhibit two forms of photoluminescence: intrinsic and extrinsic.

2.1 *Intrinsic PL*

Intrinsic photoluminescence is observed in highly pure semiconductors where radiative band-to-band transitions occur, typically at relatively high temperatures. Here, a photon excites an electron into the conduction band (CB), leaving a hole in the valence band (VB). As a result, the electron and hole recombine radiatively, emitting light. In indirect band-gap semiconductors such as aluminum arsenide (AlAs) and gallium phosphide (GaP), a phonon is required to facilitate the band-to-band transition and conserve momentum. In contrast, direct band-gap semiconductors like gallium arsenide (GaAs) do not require phonon assistance for radiative transitions, leading to a significantly higher quantum efficiency (QE) or probability of radiative recombination. This property makes direct band-gap semiconductors ideal for applications in light-emitting diodes (LEDs) and semiconductor lasers due to their superior QE. Additionally, in intrinsic semiconductors, the most intense light emission occurs at low temperatures. At these temperatures, an excited electron–hole pair can form a free exciton (FE) through Coulomb attraction, rather than undergoing a simple band-to-band transition. As a result, emission from FEs is strongest at lower temperatures [8] (Fig. 3d).

2.2 *Extrinsic PL*

It occurs through interactions involving electrons and holes at impurity states. Specifically, transitions that involve an electron that is captured by a donor, resulting in a neutral donor state (ND) (Fig. 3b) and a hole at the top of the VB, as well as between an electron at the lower edge of the CB and a hole in the neutral acceptor state (NA) (Fig. 3c). Many other excitonic recombination mechanisms occur in addition to photoluminescence (PL) from FEs. These include extremely frequent excitonic molecules at low temperatures when the state occupancy becomes high, excitonic

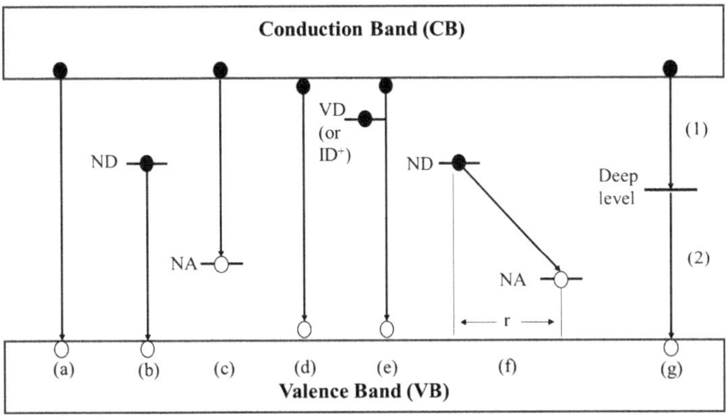

Fig. 3 The radiative recombination process in semiconductors includes: **a** band-to-band recombination; **b** transition from a neutral donor (ND) to the valence band (VB); **c** transition from the conduction band (CB) to a neutral acceptor (NA); **d** radiative recombination of free excitons (FE); **e** radiative recombination of bound excitons (BE); **f** donor–acceptor pair (DAP) recombination with a separation distance, r; and **g** deep-level luminescence defects, where either of two transitions (1) or (2) may be radiative

polarons, self-trapped excitons, and radiative recombination of bound excitons (BEs). BEs are formed when a free exciton is captured by a defect (Fig. 3e), which could be an ND, an NA, an ionized donor (ID), or an ionized acceptor (IA).

In materials with lower purity, BE recombination can become much more efficient than FE recombination, leading to a distinct peak in the highly resolved photoluminescence (PL) spectrum at lower temperatures. This makes BE recombination a sensitive method for identifying defects and impurities. Both FE and BE lines are often accompanied by optical phonon replicas, particularly in indirect-gap semiconductors like Si, Ge, and GaP, due to excitonic recombination processes. In addition, the role of excitonic recombination in nano-sized semiconductor structures, such as quantum wells, quantum wires, and quantum dots, has become more prominent, as the in-plane confinement increases the binding energy of excitons. These structures exhibit excitonic photoluminescence, even at much higher temperatures. Radiative tunneling aids in the recombination process of the donor-to-acceptor pair (DAP) when both kinds of impurities are present in the form of compensated semiconductors, as observed (Fig. 3f), where the PL spectra sporadically reflect the DAP separation [7].

2.3 *How Scattering Is Related to PL?*

Understanding the behavior of incident light when it interacts with a material involves not only knowing how much of the light is absorbed but also analyzing the components of the light that are reflected and scattered. Scattering i.e., Rayleigh Scattering (elastic with no energy loss) and Raman Scattering (inelastic with energy loss). The Raman scattering process is a two-photon interaction, where photon absorption and emission take place via a virtual intermediate state, and the re-emission of scattered light happens in picoseconds. Since Raman scattering provides information about molecular vibrations or phonons in solids and the crystallinity of materials, it is valuable for vibrational spectroscopy. However, its cross-section and quantum efficiency are low, ranging from 10^{-8} to 10^{-6}. In contrast, photoluminescence (PL) is a two-step, one-photon process, involving photon absorption followed by photon emission, with a transition time greater than 0.1 ns [7].

2.4 Fluorescence and Phosphorescence Correlation

Fluorescence is a radiative transition between states with the same spin multiplicity (singlet to singlet), making it a spin-allowed process where the system transitions from an excited state to the ground state. In contrast, phosphorescence is a spin-forbidden transition, occurring when the excited state has a different spin multiplicity (triplet) compared to the ground state. Spin-forbidden transitions happen through intersystem crossover because of the usually weak spin–orbit coupling (Fig. 1). Consequently, the phosphorescence lifespan also referred to as the radiative recombination time ranges from 1 ms. To 10 s, which is substantially longer than the fluorescence lifetime, which ranges from 0.1 to 10 ns. Since delayed fluorescence is often referred to as phosphorescence, the distinction between fluorescence and phosphorescence can sometimes be ambiguous. Metal–ligand complexes (MLCs) typically exhibit intense phosphorescence due to their significant intersystem crossing, which results from spin–orbit coupling with heavy metals. These complexes generally consist of a central transition metal atom bonded to one or more organic ligands, contributing to their strong phosphorescent properties [7].

3 What Influences Photoluminescent Properties?

3.1 Structure and Composition of Material

Materials that are synthesized in the lab exhibit different structural and compositional properties. There can be defects, impurities, or crystallinity significances that can influence PL intensity and its wavelength. Many materials like metal oxides are doped using transition elements, and the introduction of these substances into the material can create new energy states that can affect PL emissions. Different types of materials like semiconductors, nanomaterials, organic molecules, and hybrid perovskites each exhibit different PL characteristics.

3.2 Excitation Wavelength and Intensity

The excitation of photons depends upon the wavelength of the ionization sources and different wavelengths can excite photons at different electronic transitions altering the PL spectrum.

3.3 Mechanism of Energy Transfer

Forster resonance energy transfer (FRET) is a non-radiative energy transfer process between two light-sensitive molecules (chromophores), while Dexter energy transfer (DET) involves the transfer of excitation from the donor to the acceptor through electron hopping. For DET to occur, the wavefunctions of both the donor and acceptor must overlap. Consequently, electron exchange interactions between adjacent fluorophores can either quench or enhance photoluminescence (PL), potentially altering their non radiative lifetime, particularly when the fluorophores are in close proximity.

3.4 Temperature and Environment of Surrounding Medium

Increased temperature can cause non-radiative recombination as well as affect phonon populations which can influence radiative and non-radiative processes that alter PL spectrum. Oxygen, humidity, refractive index, and dielectric constant can affect light absorption and emission altering PL spectra.

4 Photoluminescence Techniques and Instrumentation

4.1 Steady-State Photoluminescence (SSPL)

SSPL (Fig. 4) is an analytical technique for the study of material properties through the measurement of light emitted due to their excitation through a continuous ionization source. A sample in steady-state PL spectroscopy is continuously excited by the use of an ionization source, usually a laser or a xenon lamp for photoluminescence and a xenon flash lamp for phosphorescence, which energizes electrons from the ground state to an excited state. As these electrons fall back to the ground state, they release photons, which are subsequently detected and analyzed. For the detection three detectors are used that allow various measurements in the UV–VIS–NIR region of ~ 200 to ~ 3000 nm.

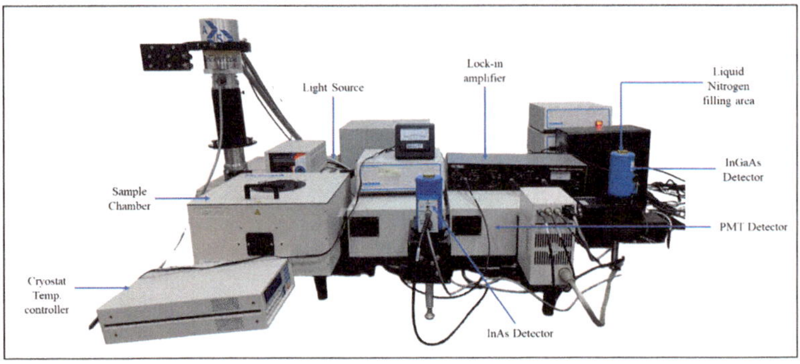

Fig. 4 Horiba SSPL instrument

4.1.1 What Type of Samples, and Sample Holders Are Used?

Samples (non-hazardous type) which are required in SSPL instrument for measurements can be in solid form, liquid form, or thin film (composed of glass or quartz) coated with the sample, and various types of sample holders are used based on the sample prepared. Low-temperature measurements of the sample are carried out in the closed-cycle helium optical cryostat (Fig. 5) allowing the spectrofluorometer to measure within a temperature range of approximately 10–400 K, and the sample must be either a thin film or a solid pallet with minimum lateral dimensions of 20 mm × 20 mm. For liquid samples, samples are kept in the cuvette (made of quartz) and placed in the liquid sample holder (Fig. 8). Powder samples and samples coated on thin film are placed in the sample holder (Fig. 6). For the measurement of quantum yield both for powder, and thin film samples, different types of frames are used in which samples are kept and placed on the PLQY sample holder, and for liquid samples direct cuvette is placed onto the PLQY sample holder, and the sample holder is covered with integrating sphere (Fig. 7).

4.1.2 Detector Selection Before Taking PL Measurement

Selecting the appropriate detector before measurements in the SSPL instrument is crucial, as it ensures accurate recording of PL spectra within the specific wavelength range of interest. Single-channel detectors (SCDs) or multichannel detectors (MCDs) are the optical detectors that are used in spectrophotometers. SCDs and MCDs are differentiated based on their active sensing elements (ASE), as ASE acts as a transducer that converts the absorbed photons that are operated within the detector's operational wavelength range into an electrical signal. Similarly, SCDs contain only one ASE that can have either an analog output (voltage/current) or a digital output (pulse counting), and MCDs contains multiple ASE that allow them to collect several wavelengths simultaneously at the focus plane of the spectrograph, with no requirement of exit slit. The selection of detector depends upon the application requirements

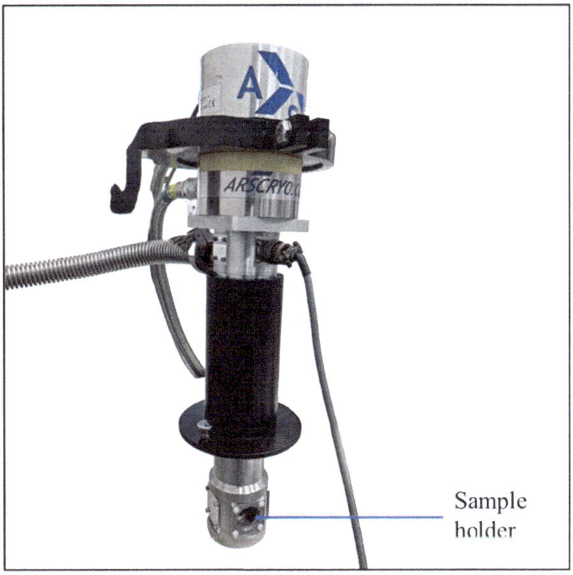

Fig. 5 Helium optical cryostat sample holder

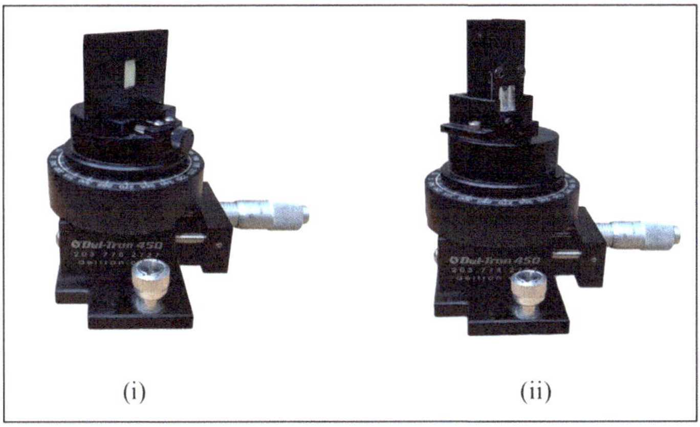

Fig. 6 Sample holders which contain sample. **a** Sample coated in thin film; **b** sample in powder form

as each detector type has its advantages and disadvantages. In the SSPL instrument, the detection system comprises three types of detectors (Fig. 4): a photomultiplier tube (PMT) for recording spectra in the wavelength range of 200–850 nm, an Indium Gallium Arsenide (InGaAs) detector for spectra from 800 to 1550 nm, and an Indium Arsenide (InAs) detector for spectra spanning 1000–3000 nm. Detectors like InGaAs and InAs liquid nitrogen are added before taking the measurements. Once the detector is cooled below the room temperature, the background signal as well as the noise associated with the component used in SSPL are reduced.

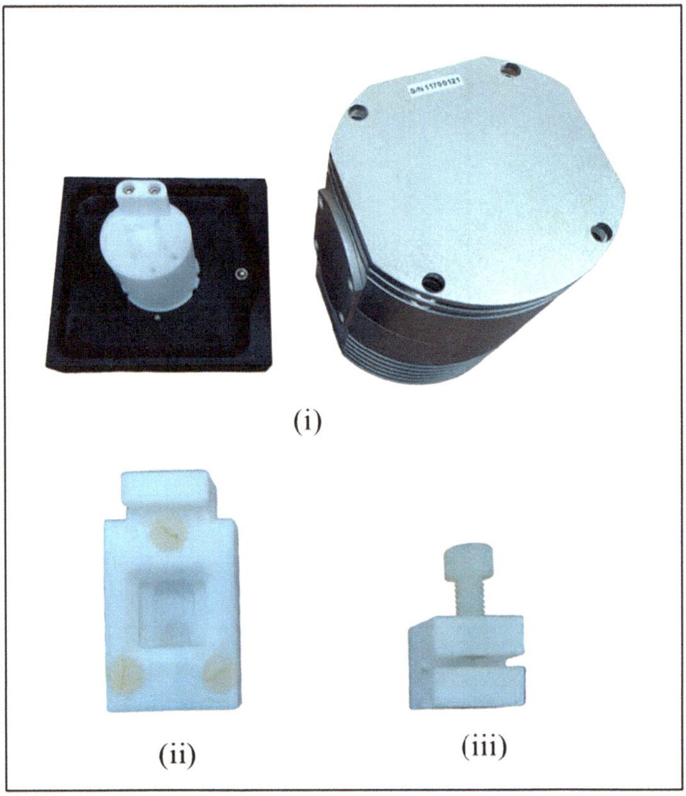

Fig. 7 **a** PLQY sample holder with integrating sphere; **b** sample frame for keeping powder samples; **c** sample frame for keeping thin film samples

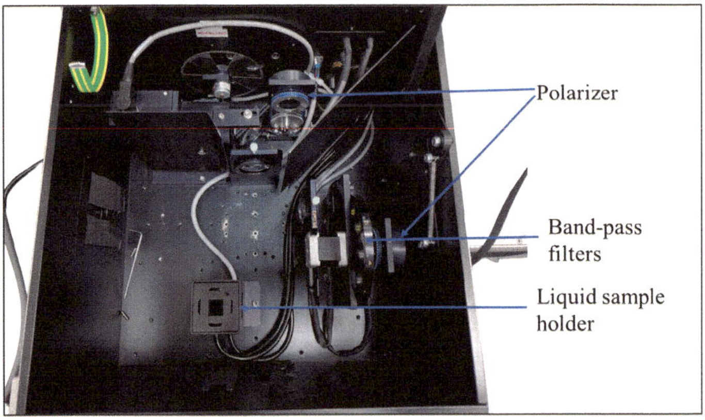

Fig. 8 SSPL instrument sample chamber having polarizer that needs to be aligned in the path of emission and excitation light source. Also, it has liquid sample holder where cuvette is placed, and the band-pass filters for signal suppression

4.1.3 Signal Suppression

A filter is a device used to suppress unwanted components or characteristics from a signal, either entirely or partially, by blocking specific wavelengths. This is commonly done to minimize background noise and eliminate interference by removing certain frequencies. Filters come in various types: a Low-pass filter allows low frequencies to pass while attenuating higher ones, a High-pass filter allows high frequencies to pass while attenuating lower ones, a Band-pass filter allows only frequencies within a specific range to pass, and a Band-stop filter attenuates frequencies within a certain range. In SSPL instruments, band-pass filters are typically employed (Fig. 8) [9].

4.1.4 Type of Measurements

There is a finite number of vibration frequencies that molecules can have, and resonance can arise when a molecule's frequency matches that of a light wave, in that case, the molecule is said to absorb light. A luminous material exhibits its characteristic wavelength at which maximum excitation and emission take place when it absorbs light.

4.1.5 Emission Spectra

Emission spectra refer to the specific frequencies of electromagnetic radiation emitted by a substance as it transitions from a high-energy state to a lower one. This occurs when atoms or molecules in an excited state release energy as photons, producing distinct lines or bands that are characteristic of the elements or compounds involved. In emission energy spectra (Em_{sp}) the monochromator of excitation photon energy (Ex_{pe}) is monitored at a fixed wavelength while the monochromator of the emission photon energy (Em_{pe}) is varied until the PL spectra are obtained at the maximum Ex_{pe} corresponding to the excitation energy.

4.1.6 Excitation Spectra

An excitation energy spectrum (Ex_{sp}) refers to a graph that plots the intensity of emitted light, either fluorescence or phosphorescence, against the excitation wavelength. Such a spectrum is obtained when the monochromator of the Em_{pe} is monitored at a fixed wavelength while the monochromator of the Ex_{pe} wavelength is changed progressively, and the excitation energy spectra are obtained. Ex_{sp} gives information regarding the range of wavelengths that are best for exciting a sample, and normally it closely resembles the absorption spectrum of the material under investigation.

4.1.7 Quantum Yield and Quantum (Energy) Efficiency

Rubin and Braslavsky have detailed the history of the term 'quantum yield' and the nomenclature used before its initial application in 1925. According to the IUPAC definition, the integral quantum yield, $\Phi\lambda$, is [10, 11]:

$$\Phi\lambda = \frac{\text{number of event}}{\text{number of photons absorbed}}$$

where, $\Phi\lambda$ can be used for photophysical processes (such as, e.g., intersystem crossing, luminescence, ejection of electrons in the case of photomultipliers, or photochemical reactions). More usually, it is written for the emission of light [10, 12]:

$$\Phi\lambda = \frac{\text{number of photons emitted}}{\text{number of photons absorbed}}$$

The wavelength λ refers to the photons absorbed. When reporting quantum yield data, it is crucial to also specify other experimental conditions such as temperature, sample form, irradiance, and the emission wavelength range observed. Since the equation provided does not account for time, the quantum yield indicates the relative contributions of radiative and non-radiative processes within the chromophore [10, 13]. While measuring the Quantum yield of the sample, first the PL spectra of the substrate or the matrix are to be recorded and then the PL spectra of the sample are recorded keeping the monochromator of the Ex_{pe} at fixed wavelength and varying the wavelength of the Em_{pe} accordingly. To measure photoluminescence (PL), the spectra of fluorescence emission (Fe) and scattering (Fs) from both the sample and the matrix are recorded. By analyzing these spectral measurements, the photoluminescence quantum yield (PLQY) can be determined using the following equation:

$$Q_F = Q_R \frac{I_F}{I_R} \frac{A_R}{A_F}$$

where,

Q_F quantum yield of the unknown fluorescent sample,
Q_R quantum yield of the reference standard,
I_F integrated fluorescence intensities for the unknown,
I_R integrated fluorescence intensities for the reference,
A_F absorbance values of the unknown, and
A_R absorbance values of the reference.

4.1.8 Fluorescence Decay Spectra (or Lifetime)

FLT, or Fluorescence Lifetime, refers to the time a fluorophore stays in its excited state before emitting a photon and returning to its ground state. This duration can range from picoseconds to hundreds of nanoseconds, depending on the specific fluorophore. The FLT of a group of fluorophores is the time it takes for the number of excited molecules to decay exponentially to N/e, or 36.8% of the original population, through energy loss via fluorescence or non-radiative processes. FLT is an inherent property of a fluorophore, unaffected by factors such as fluorophore concentration, sample absorption, sample thickness, measurement technique, fluorescence intensity, photobleaching, and excitation intensity. However, it is influenced by external factors like temperature, polarity, and the presence of fluorescence quenchers, while internal factors are related to the structure of the fluorophore [14]. Fluorescence lifetime can be measured using either the frequency-time or time domain methods. In SSPL, the time-domain approach is employed, where an ionization source is used to excite a sample, which could be a cuvette, powder, or thin film. The emission intensity is then recorded over time. The FLT is determined from the slope of the resultant decay curve shown in the figure. When the PLQY measurement is taken in the SSPL instrument few parameters should be known before taking the measurements. Parameters like: monochromator wavelength of the Ex_{pe} and Em_{pe}; start and end time of the measurement; shorts (number of photons that are counted on the least count of the detector); and lamp frequency. Thus, fluorescence lifetime is a very important photophysical parameter, which provides insights regarding the information on the energy relaxation and dynamics of the studied species, including molecular state energy transfer, molecular rotation, and dynamic quenching.

4.1.9 Steady-State Fluorescence Anisotropy or Fluorescence Polarisation

Steady-state fluorescence anisotropy, or fluorescence polarization, refers to the changes in a molecule's orientation in space during the interval between absorption and emission. It characterizes how the molecule's dipoles are aligned relative to the electric field vectors of both the excitation and emitted light. When a group of fluorophores is excited with vertically polarized light, the emitted light undergoes partial depolarization, which depends on the rate of molecular rotation in the solution. Faster molecular rotation leads to greater depolarization of the emitted light, while slower motion results in the emitted light maintaining more of the polarization from the initial excitation [15]. Polarization provides significant flexibility in optical measurements. For SSPL spectrophotometry, a linearly polarized laser or polarizer is required in the path after the ionization source or excitation monochromator, and an analyzer must be positioned in front of the emission monochromator on the detection path, as shown in Fig. 8.

4.2 *Time-Resolved Photoluminescence (TRPL)*

Time-Resolved Photoluminescence (TRPL), also referred to as Time-Correlated Single Photon Counting (TCSPC), is a method that aligns the timing of a pulsed excitation source, such as a laser or LED, with the detection of individual photons. This synchronization enables the reconstruction of fluorescence lifetime decay over many repeated excitation pulses and photon detections. The TRPL decay profile, shaped by various competing radiative and non-radiative recombination processes in the system and its surroundings, offers valuable information about these processes. Ideally, only 0 or 1 photon is detected after each excitation pulse, requiring millions to billions of excitation pulses to achieve a high signal-to-noise ratio [16]. The TRPL intensity I_{TRPL} measured after P pulses is thus the sum of the TRPL intensity I^{p}_{TRPL} after each excitation pulse P [17, 18]:

$$I^{(t)}_{TRPL} = \sum_{p=1}^{p} I^{p}_{TRPL}(t)$$

TRPL decays have been classically fitted with a sum of exponential decays $A_i e^{-t/\tau_i}$ (A_i and τ_i are respectively the amplitude and lifetime of component i) [19]:

$$I^{p}_{TRPL} = \sum_{i}^{n} A_i e^{-t/\tau_i}$$

4.2.1 Working Principle of TRPL

TRPL is particularly well-suited for measuring short photoluminescence decay times, ranging from sub-nanoseconds to hundreds of nanoseconds, using very low excitation intensities to prevent nonlinear effects or potential damage to the sample. Block schematic diagram (Fig. 9) of the working principle of TRPL; here, normally, a mode-locked laser is used for the ionization source. The light pulse splits into start and stop channels. In the starting channel, then, a trigger (start) pulse is created either synchronously by the electronic driver that fires (the discharge of) the ionization source, or by a PMT or APD monitoring the light pulse divided by a beamsplitter. The events in TRPL follows [20];

i. The start pulse is generated either by a Constant Fraction Discriminator (CFD), which reduces the effects of variations in pulse height, or by a Leading-Edge Discriminator (LED), particularly when using a stable laser system that delivers light pulses with uniform height. This start pulse subsequently activates a Time-to-Amplitude Converter (TAC), initiating a voltage ramp.
ii. At the stop channel, the light pulse excites the sample in photoemission, and, after an arbitrary time-lapse denoted as t, the sample emits a single photon.

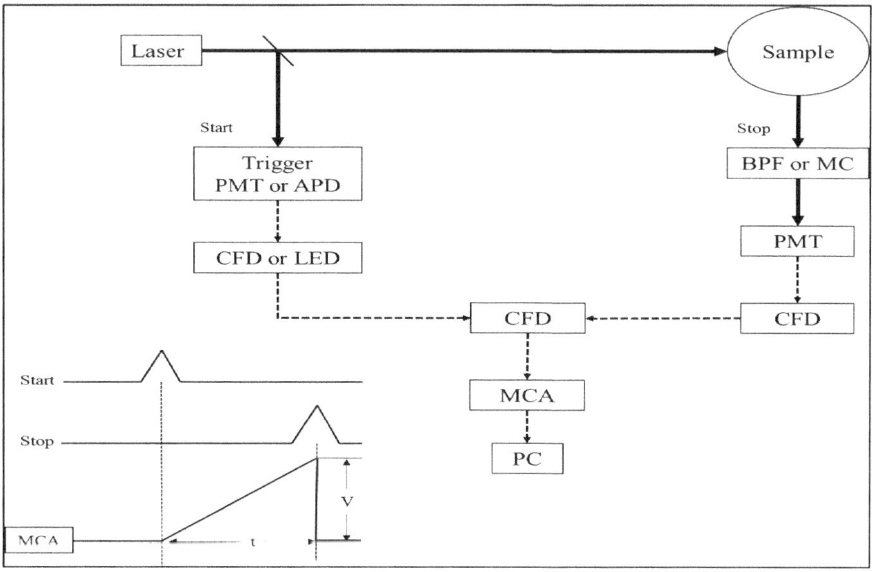

Fig. 9 The block diagram of a conventional TRPL system includes various components: an avalanche photodiode (APD), a band-pass filter (BPF), a monochromator (MC), a time-to-amplitude converter (TAC), a multichannel analyzer (MCA), a constant fraction discriminator (CFD), and a personal computer (PC). In the schematic, solid lines represent optical pathways, while dotted lines indicate electrical pathways. The setup features key elements such as photomultiplier tubes (PMTs), APDs, constant fraction discriminators, light-emitting diodes (LEDs), and leading-edge discriminators

iii. Just before the photo-multiplier tube (PMT), a band-pass filter (BPF) or a monochromator (MC) is installed. The light power is thus dumped to very low levels, and at most one photon is intended for detection with the PMT at every photoexcitation event.
iv. The Constant Fraction Discriminator (CFD) eliminates noise pulses from the single-photon events detected by the photomultiplier tube (PMT) in the stop channel. It then stops the Time-to-Amplitude Converter (TAC), which generates a ramp voltage V, directly proportional to the photon detection time t, between the start and stop pulses and sends it to the multichannel analyzer.
v. Each channel number in the multichannel analyzer (MCA) corresponds to a specific voltage (V) and, consequently, to a particular time (t). When numerous single-photon events are recorded, the count of photoemission occurrences in each channel creates a histogram, which serves as a representation of the photoluminescence (PL) decay curve.

Note: Resetting the Time-to-Amplitude Converter (TAC) by discharging its capacitor takes a few microseconds, leading to a period of system dead time. This can be mitigated by switching operational modes, where the photomultiplier tube (PMT) pulse corresponding to single-photon emission is directed to the TAC's start input, and a trigger pulse synchronized with photoexcitation stops the TAC. This approach shortens measurement time since the TAC only activates when a single photon is emitted, unlike the standard mode where it starts with every cycle. This method enhances time resolution significantly compared to traditional techniques, as the accuracy of time measurements is no longer constrained by the width of detector pulses.

This technique relies on the principle that the detection of a single photon at a time, t after an excitation pulse is proportional to the fluorescence intensity at that time. It involves repeating laser or LED pulses at a high repetition rate, usually ranging from 10 kHz to 100 MHz, and synchronizing these pulses with the arrival time of the subsequent photon at a detector, such as a photomultiplier tube (PMT). The timing electronics then record these events in sequence with either a time-to-digital converter (TDC) or a time-to-amplitude converter (TAC). This process is continued until sufficient data has been obtained to reconstruct the decay curve accurately, and the resultant decay still exhibits an exponential behavior in determining the lifetime decay, *t*.

4.2.2 How Does TRPL Analyse Quenching?

Quenching is the process by which fluorescence is reduced because of an increase in the nonradiative decay rate (K_{nr}). The phenomenon of quenching may be broadly classified into two types: static and dynamic. According to static quenching, the fluorophore forms a nonfluorescent complex with a quencher, leading to decreased fluorescence emission. Static quenching occurs before the event of excitation and does not have any effect on the fluorescence lifetime of remaining fluorescent molecules. So, the net results in a decrease in emission intensity because some fractions of the fluorophores are complexed into a non-fluorescent species. Dynamic quenching, also referred to as collisional quenching, excited fluorophore interactions with quencher molecules take place after the fluorophore has been excited. Such interactions led to an increase in the nonradiative decay (K_{nr}) rate, thus causing a decrease in fluorescence intensity and lifetime [21].

Dynamic quenching differs from static quenching (Fig. 10) in its direct impact on the excited-state dynamics of fluorophores. Emission intensity will decrease in both forms of quenching; however, an evident change in fluorescence lifetime occurs in only dynamic quenching. This is important for the distinction or interpretation of fluorescence data regarding the type of quenching mechanism involved. Provided that the kinetics are of the Stern–Volmer type, fluorescence quenching is then linear for quencher concentration. Quenching data are usually analyzed using the so-called Stern–Volmer equation [22]:

$$\frac{Io}{I} = 1 + K_{sv}\left[Q\right]$$

where

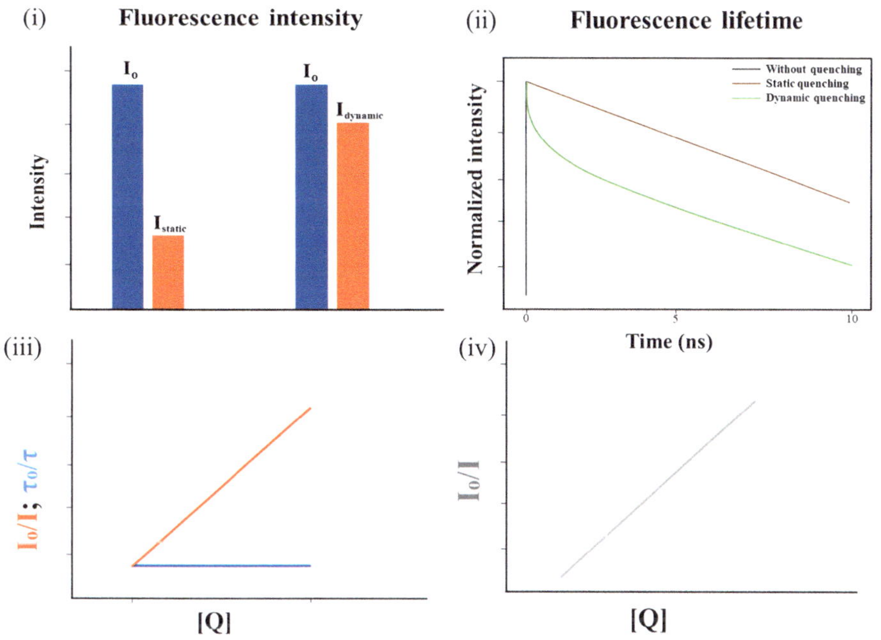

Fig. 10 Difference in static and dynamic quenching; **a** TRPL; **b** Stern–Volmer plot; **c** static; and **d** dynamic quenching

Io fluorescence intensity in the absence of a quencher,
I fluorescence intensity in the presence of a quencher,
K_{sv} Stern–Volmer quenching constant, and
[Q] concentration of the quencher.

If an average fluorescence lifetime is used in the analysis, it should be calculated using the intensity-weighted average to account for the contributions of different fluorophores or different quenching mechanisms present in the sample. The intensity-weighted average lifetime (τ_{avg}) is given by:

$$\tau_{avg} = \frac{\sum (I_i \cdot \tau_i)}{\sum I_i}$$

where,

I_i intensity,
τ_i lifetime of each component.

This approach ensures that the average lifetime accurately reflects the overall fluorescence behavior of the sample.

4.2.3 Role of Photobleaching in TRPL Measurement

Photobleaching is a process whereby fluorophores no longer fluoresce after some time of exposure to the excitation ionization source. The role of this phenomenon is that, with time, fewer active fluorophores exist, thus lowering the general fluorescence emission intensity. With increased time, the run times of measurements necessarily have to be longer to collect enough data for analysis since the reduced emission intensity makes it harder to get a clear signal. It is important to point out, however, that photobleaching only affects the number of fluorophores and their emission intensity and does not alter their fluorescence lifetime. Fluorescence lifetime is an inherent characteristic of the fluorophore and refers to the average time a molecule spends in its excited state before returning to its ground state through the process of emitting a photon. This value will remain constant despite any change in the concentration or number of fluorophores. Therefore, while photobleaching has practical effects (Fig. 11) for data collection, it has no fundamental impact on the fluorescence lifetime of interest.

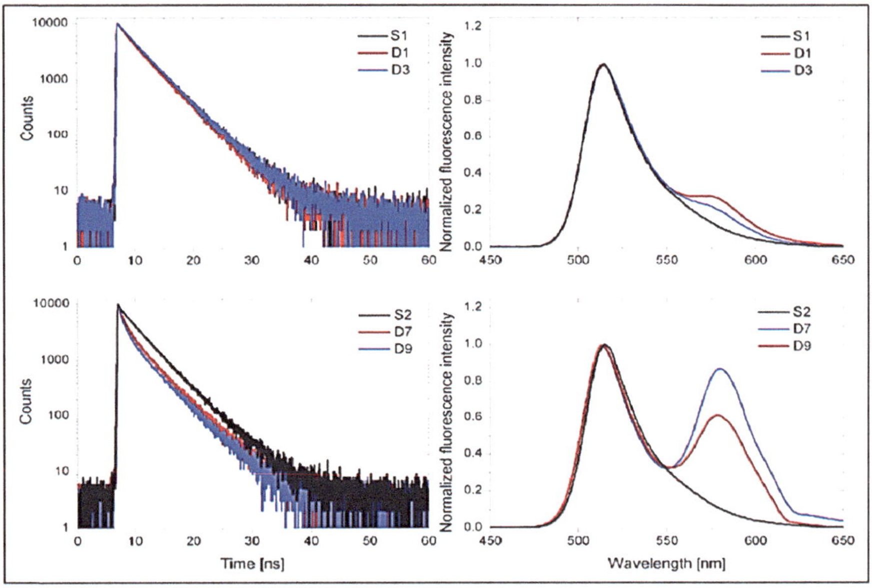

Fig. 11 The fluorescence lifetime and normalized fluorescence intensity. Example of photobleaching of the sample obvious in intensity measurement

5 Advances in PL Instrumentation and Technology

5.1 Integration with Raman Spectroscopy

Photoluminescence spectroscopy is a tool for the investigation of the composition and solid-state structure of materials, and the high spectral resolution provided by the Raman spectrophotometer intensifies the capability of the photoluminescence spectroscopy, especially in cases of narrow emission bands or lines. The possibility of performing both photoluminescence and Raman spectroscopy in the same instrument makes it very suitable for studying 2D crystals. In such applications, the Raman spectrometer is a twofold instrument that will help get all information about the material's properties. Obtaining a Raman spectrum with low-concentration samples is very challenging due to the weak nature of Raman scattering. Also, a strong fluorescent background and detector noise overwhelms Raman signals therefore, performing laser-excited PL with a Raman spectrophotometer is ideal [23].

Aluminum oxide (Al_2O_3), often called corundum in its single crystal form, as Al-O bonds are ~ 60% ionic and ~ 40% covalent. In corundum, aluminum atoms are in the + 3 formal oxidation state, surrounded by six oxygen atoms each in the − 2 formal oxidation state to form a distorted octahedron. Pure corundum is colourless, but transition metal impurities, even at levels below 1%, can color it. For example, Cr^{3+} impurities give the red color of ruby, and titanium impurities color sapphire. The electron configuration of the Cr atom is [Ar] $3d^5 4s^1$, and when occupying an Al site in the corundum crystal lattice Cr^{3+} has the electron configuration [Ar] $3d^3$. The absorption and emission transitions associated with the 3d orbitals of the t_{2g-eg} transitions are influenced by the host lattice crystal field [24]. Raman and photoluminescence spectroscopies together can be used to characterize a host crystal and its transition metal impurities.

5.2 Integration with Electron Microscopy

Quantum dots synthesis aims to produce a homogenous batch of particles to attain uniformity in shape, size, and composition. However, there are some batches of particles where some degree of heterogeneity is produced, both in structure and optical properties. A novel method combining cryogenic correlative single-particle photoluminescence (PL) spectroscopy with electron tomography enables the analysis of photophysical and structural variations in quantum dots. This technique captures the photoluminescence characteristics of individual quantum dots and correlates them with their three-dimensional structures, obtained via cryogenic electron tomography. This approach sheds light on the electronic configuration, physical structure, and surrounding environment of single quantum dots, uncovering heterogeneities that are typically lost in ensemble-averaged measurements but preserved in single-particle spectroscopy. Transmission electron microscopy (TEM) and optical spectroscopies, such as cathodoluminescence (CL)—the emission of light when a material is exposed to an electron beam—and photoluminescence, are extensively employed to investigate the structural, optical, and electronic properties of lattice defects in semiconductor materials. TEM is commonly used to study larger defects, including their characteristics, distribution, chemical composition, and atomic structures, while optical spectroscopy focuses on characterizing point-defect complexes. Electron irradiation can sometimes induce lattice defects, thereby altering the electronic properties of materials either temporarily or permanently. Furthermore, electron excitation triggers radiative electronic transitions and various non-radiative recombination processes. When examined using TEM, a specimen often exhibits CL emission. To differentiate PL emission from CL, PL spectra are recorded with the electron beam turned off [27].

6 Applications of Photoluminescence in Material Science

6.1 Semiconductors

Photoluminescence (PL) in semiconductors is a widely used method for material characterization, typically examined through spectroscopy techniques that measure the energy distribution of photons emitted after optical excitation. This energy distribution provides valuable insights into material properties, including defect types and concentrations, as well as stimulated emission—where photons of a specific frequency induce an excited atom to transition to a lower energy state. PL spectroscopy is utilized to evaluate optical emission efficiency, analyze material composition (such as alloy content), identify impurities, and determine layer thicknesses (like quantum well dimensions). The discussion of PL emission in semiconductor nanostructures, particularly those utilizing the GaAs/AlGaAs heterostructure system, emphasizes phenomena such as quantum confinement, dimensionality reduction, and atomic-scale structural influences. This technique generates spectral data that is essential for understanding various recombination processes, including intrinsic mechanisms like excitonic recombination and electron–hole liquid recombination, defect-related recombination, hot-carrier recombination, and recombination facilitated by phonons [28].

6.2 Nanomaterials

PL in the field of nanomaterials, providing valuable insights into the optical and electronic properties of nanomaterials. PL helps in the characterization and optimization of quantum dots, nanowires, carbon nanotubes, and other nanomaterials that are widely applied in the diverse fields of sensing and light-emitting technologies because the size dependence of the nanomaterials strongly influences the optical properties related to PL emission. In particular, the size of nanomaterials may result in changes to optical behavior such as emission wavelength and intensity. Besides, surface states play a major role in defining the PL characteristics of nanomaterials. For instance, surface traps may quench PL or give new features of emission that have been shown very useful for getting information about the quality and stability of the surface. In this direction, modification or functionalization of the nanomaterial surface also affects PL properties. Tailoring of nanomaterials for applications such as sensors or catalysts can be attained in these effects, depending on the performance criterion desired.

6.3 Organic Molecules and Polymers

PL techniques are applied to the study of organic materials to enhance their emission properties and, correspondingly, optimize their performance in OLEDs for more efficient and brighter display and lighting technologies. PL is used to analyze photoexcitation and charge transfer processes in organic photovoltaic materials. This helps to enhance the efficiency and stability of OPVs and makes them more viable for applications in solar energy.

6.4 Hybrid Perovskites

PL is one of the techniques for the investigation of charge carrier dynamics, defect states, and the overall efficiency in hybrid perovskite (it is a naturally occurring calcium titanium oxide mineral and a class of compounds with a similar crystal structure) materials. The information obtained will be critical to optimizing perovskite solar cells to realize higher power conversion efficiency with improved stability. This knowledge can serve as a good platform for the development of high-performance light-emitting perovskite devices, including both LEDs and lasers representing bright and tuneable emissions [29].

7 Industrial Applications of Photoluminescence

7.1 Biomedical Field

Among nanomaterials, photoluminescent nanoparticles, such as quantum dots, gold nanoparticles, and lanthanide-based upconverting nanoparticles, have the dual advantages of therapeutic effects and bioimaging capability. The unique optical properties make these two categories of nanomaterials very promising in both optical imaging

(OI) and the initiation of photochemical reactions in biological settings. The advancement of photoluminescent nanoparticles as contrast agents or theragnostic tools offers the potential for precise analysis of specific biological processes and targeted disease treatments. Researchers are currently exploring innovative nanoscale materials that provide enhanced detection sensitivity, multimodal imaging functionality, and superior therapeutic efficacy. Notable examples include persistent luminescent nanophosphors and carbon dots (C-dots), which have shown significant promise as alternatives to conventional luminescent nanoprobes [30].

7.2 Environmental Monitoring

The sensitivity and specificity that photoluminescence offers, it finds significant applications in different environmental applications, such as detection and monitoring of pollutants, environmental change assessment, and enhancement of sustainable practices. PL sensors can trace heavy metals like mercury, lead, and cadmium in water, which helps in pollution control and management of the quality of water. PL-based techniques can detect organic pollutants, pesticides, herbicides, and other industrial chemicals, which become very important in environmental monitoring. Photoluminescent materials can enable the detection of particulate matter and aerosols in the air, possibly supplying information for air quality assessment and strategies to mitigate pollution. PL sensors can track harmful gases like NO_x, SO_x, and CO_2 and hence help control their emissions to achieve better air quality. Those photoluminescent probes are devices that can measure parameters such as pH, dissolved oxygen, and specific ions or compounds in solution to monitor water quality. PL techniques can monitor soil health by detecting contaminants and nutrients, thus sustaining agriculture and land management practices. Photoluminescent nanomaterials can be employed in remediation efforts at contaminated sites to offer efficient, targeted approaches to cleaning up pollutants. To this end, PL-based photocatalysts can degrade organic pollutants in water and air, promoting environmental sustainability through advanced oxidation processes [31].

7.3 Pharmaceutical Industry

PL techniques can be utilized for the tracking and monitoring of the sites of drug distribution by the drug-loaded nanocarriers in biological systems to ensure the delivery of the correct amount with precision to target locations. PL could also help study the kinetics of the release of a drug from different delivery systems, thus allowing for the development of control release formulations for better therapy. On the other hand, photoluminescent nanoparticles, such as quantum dots, play an integral role in theragnostic applications, whereby therapy and diagnostics merge to trace back treatment efficiency in real-time. PL spectroscopy is applied to the investigation of binding interactions of a drug with its target protein in the design of more effective pharmaceuticals. PL methods can monitor activities and inhibition of enzymes involved in metabolic pathways, providing very useful information about potential targets of drugs. Photoluminescence is applied to investigate the interaction of different drugs with nucleic acids in antiviral and anticancer therapies. By applying these special properties of photoluminescence, the pharmaceutical industry will be able to achieve the goals of more efficient drug delivery, enhanced biomolecular interactions, and better imaging capabilities for eventual improvement in therapeutic strategies and the quality of life [32].

References

1. Lastusaari, M., Laamanen, T., Malkamäki, M., Eskola, K.O., Kotlov, A., Carlson, S., et al.: The Bologna Stone: history's first persistent luminescent material. Eur. J. Mineral. **24**, 885–890 (2012)
2. Stokes, G.G.: O promjeni refraktivnosti svjetlosti. Philos. Trans. R. Soc. Lond. **142**, 263–562 (1852)
3. Sekiya, M., Yamasaki, M.: Antoine Henri Becquerel (1852–1908): a scientist who endeavored to discover natural radioactivity. Radiol. Phys. Technol. **8**, 1–3 (2015)
4. Berberan-Santos, M.N.: Pioneering contributions of Jean and Francis Perrin. Science **19**, 11 (1979)
5. Valeur, B., Berberan-Santos, M.N.: A brief history of fluorescence and phosphorescence before the emergence of quantum theory. J. Chem. Educ. **88**, 731–738 (2011)
6. Valeur, B.: From well-known to underrated applications of fluorescence. In: Fluorescence of Supermolecules, Polymers, and Nanosystems, pp. 21–43 (2007)
7. Aoki, T.: Photoluminescence spectroscopy. In: Characterization of Materials, pp. 1–12. Wiley (2012)
8. Karaiskaj, D., Engtrakul, C., McDonald, T., Heben, M.J., Mascarenhas, A.: Intrinsic and extrinsic effects in the temperature-dependent photoluminescence of semiconducting carbon nanotubes. Phys. Rev. Lett. **96**, 106805 (2006)
9. Foreman, P.C.: The excitation and emission spectra of fluorescent components of human dentine. Arch. Oral Biol. **25**, 641–647 (1980)
10. Wong, K.L., Bünzli, J.-C.G., Tanner, P.A.: Quantum yield and brightness. J. Lumin. **224**, 117256 (2020)
11. McNaught, A.D., Wilkinson, A.: Compendium of Chemical Terminology: IUPAC Recommendations (1997)
12. Rubin, M.B., Braslavsky, S.E.: Quantum yield: the term and the symbol. A historical search. Photochem. Photobiol. Sci. **9**, 670–674 (2010)
13. Rea, M.S.: Lighting simply made better: providing a full range of benefits without much fuss. Build. Environ. **144**, 57–65 (2018)
14. Berezin, M.Y., Achilefu, S.: Fluorescence lifetime measurements and biological imaging. Chem. Rev. **110**, 2641–2684 (2010)
15. Ba Hoang, T., Moses, A.F., Ahtapodov, L., Zhou, H., Dheeraj, D.L., Van Helvoort, A.T.J., et al.: Engineering parallel and perpendicular polarized photoluminescence from a single semiconductor nanowire by crystal phase control. Nano Lett. **10**, 2927–2933 (2010)
16. Péan, E.V., Dimitrov, S., De Castro, C.S., Davies, M.L.: Interpreting time-resolved photoluminescence of perovskite materials. Phys. Chem. Chem. Phys. **22**, 28345–28358 (2020)
17. de Melo, J., Costa, T., De Castro, C.S., Maçanita, A.L.: Photophysics of fluorescently labeled oligomers and polymers (2013)

18. Kazmerski, L.L.: Innovation in solar technology: toward a 100% renewable electricity future. In: Sustainable Energy Development and Innovation: Selected Papers from the World Renewable Energy Congress (WREC) 2020, pp. 3–12. Springer (2022)
19. Brenes, R., Guo, D., Osherov, A., Noel, N.K., Eames, C., Hutter, E.M., et al.: Metal halide perovskite polycrystalline films exhibiting properties of single crystals. Joule **1**, 155–167 (2017)
20. Becker, W.: The bh TCSPC Handbook. Becker & Hickl GmbH (2021)
21. Eftink, M.R.: Fluorescence quenching: theory and applications. In: Topics in Fluorescence Spectroscopy: Principles, pp. 53–126. Springer (2002)
22. Wahba, M.E.K., El-Enany, N., Belal, F.: Application of the Stern–Volmer equation for studying the spectrofluorimetric quenching reaction of eosin with clindamycin hydrochloride in its pure form and pharmaceutical preparations. Anal. Methods **7**, 10445–10451 (2015)
23. Tuschel, D.: Photoluminescence spectroscopy using a Raman spectrometer (2016)
24. Nassau, K.: The Physics and Chemistry of Color: The Fifteen Causes of Color (2001)
25. Moerner, W.E., Kador, L.: Optical detection and spectroscopy of single molecules in a solid. Phys. Rev. Lett. **62**, 2535 (1989)
26. Dahlberg, P.D., Perez, D., Su, Z., Chiu, W., Moerner, W.E.: Cryogenic correlative single-particle photoluminescence spectroscopy and electron tomography for investigation of nanomaterials. Angew. Chem. Int. Ed. **59**, 15642–15648 (2020)
27. Ohno, Y., Takeda, S.: A new apparatus for in situ photoluminescence spectroscopy in a transmission electron microscope. Rev. Sci. Instrum. **66**, 4866–4869 (1995)
28. Gilliland, G.D.: Photoluminescence spectroscopy of crystalline semiconductors. Mater. Sci. Eng. R. Rep. **18**, 99–399 (1997)
29. Shirota, Y., Kageyama, H.: Organic materials for optoelectronic applications: overview. In: Handbook of Organic Materials for Electronic and Photonic Devices, pp. 3–42 (2019)
30. Ibarra-Ruiz, A.M., Rodríguez Burbano, D.C., Capobianco, J.A.: Photoluminescent nanoplatforms in biomedical applications. Adv. Phys. X **1**, 194–225 (2016)
31. Reardon, K.F., Zhong, Z., Lear, K.L.: Environmental Applications of Photoluminescence-Based Biosensors, pp. 143–157 (2009)
32. Bansal, A., Zhang, Y.: Photocontrolled nanoparticle delivery systems for biomedical applications. Acc. Chem. Res. **47**, 3052–3060 (2014)

The Relationship Between Optical Imaging and Spatial Frequency

S. S. Goutam Buddha, Nagendra Kumar, Karuna S. Mallik, and Santanu Konwar

Abstract Optical imaging has long been a key method of investigating scientific inquiry, enabling visualization and analysis of structures at various scales, from the microscopic to the macroscopic. The concept of spatial frequency is central to understanding and developing optical imaging techniques. This article explores the relationship between optical imaging and spatial frequency, delving into the fundamental principles, practical implications, and advancements in imaging technology that hinge on this relationship. This paper elucidates the critical influence of spatial frequency on the capabilities and limitations of optical imaging systems.

Keywords Spatial frequency · Signal processing · Optical imaging · Diffraction limit

1 Introduction

Optical imaging encompasses a wide range of techniques used to capture images of objects using light, from conventional photography to advanced microscopy and astronomical imaging [1–7]. At the core of these imaging systems is spatial frequency, which describes how details in an image are represented in terms of sinusoidal patterns of varying intensity. Spatial frequency, measured in cycles per unit distance (e.g., cycles per millimeter), provides a framework for understanding the resolution and contrast of optical images.

S. S. Goutam Buddha (✉)
Department of Physics, Tata Institute of Fundamental Research, Hyderabad, India
e-mail: ssg.buddha@gmail.com

N. Kumar
Department of Physics, Indian Institute of Technology Delhi, New Delhi, India

K. S. Mallik
Department of Physics, Brainware University, Barasat, Kolkata, West Bengal, India

S. Konwar
Department of Physics, Abhayapuri College, Abhayapuri, Bongaigaon, Assam, India

A. Mishra and V. Pathak (eds.), *Smart Materials for Energy Storage and Biomedical Applications*, Engineering Materials, https://doi.org/10.1007/978-3-031-92584-9_11

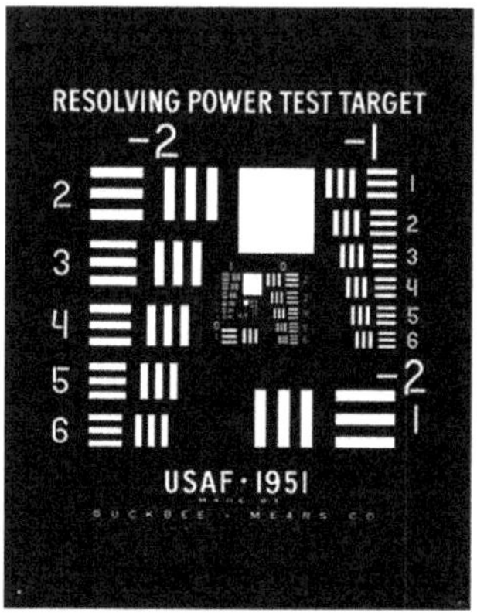

Fig. 1 Image of a "Resolving Power Test Target" from 1951, used for evaluating optical resolution. The design includes multiple groups of horizontal and vertical lines of varying thickness, labeled with numbers from 2 to 6 and −2 to −1. The pattern helps determine the resolving power of optical systems by assessing the smallest line pairs that can be distinguished

Understanding the relationship between optical imaging and spatial frequency is crucial for improving imaging technologies, particularly in fields requiring high-resolution and high-contrast images, such as medical imaging, material science, and astronomy. This article provides a comprehensive exploration of how spatial frequency interacts with optical imaging principles, influencing the quality and effectiveness of captured images. Figure 1 shows 1951 USAF, a standardized resolution chart commonly used to evaluate the optical performance and resolving power of imaging systems, which contains several horizontal and vertical lines with different line pair/mm (lp/mm).

2 The Fundamentals of Spatial Frequency

Spatial frequency, a concept rooted in signal processing, measures how frequently the intensity of a signal or image changes per unit distance. In optical imaging, it describes the repetition of features like edges, textures, or patterns across an image plane. High spatial frequencies correspond to rapid intensity changes, representing fine details such as sharp edges or intricate textures, while low spatial frequencies indicate gradual variations, capturing coarse structures like large shapes or smooth gradients. This distinction helps analyze and describe the level of detail present in an image [8–10].

The concept is central to many imaging applications. For instance, an imaging system's ability to capture high spatial frequencies in resolution analysis determines how well it can resolve fine details. Filtering techniques use spatial frequency to enhance or suppress specific features: high-pass filters emphasize fine details, while low-pass filters smooth out noise or broad structures. In medical imaging, spatial frequency analysis aids in balancing detail visibility and noise reduction. Additionally, image compression algorithms leverage spatial frequency by reducing less perceptible high-frequency components to save storage. These applications illustrate its role in understanding, optimizing, and manipulating visual data.

2.1 Mathematical Representation

Mathematically, an image can be decomposed into a series of sinusoidal functions of different spatial frequencies through a process known as Fourier analysis. Figure 2 illustrates the combination of multiple sinusoidal waves, both low and high frequency, to create an image with detailed features [9, 10].

For a continuous 2D function $f(x, y)$ representing an image, the 2D Fourier transform $F(u, v)$ is mathematically defined as:

$$F(u, v) = \int_{-\infty}^{\infty} \int_{-\infty}^{\infty} f(x, y) e^{-i2\pi(ux+vy)} dxdy$$

where u and v represent the spatial frequency variables n horizontal and vertical directions respectively, i is the imaginary unit ($i^2 = -1$).

The Fourier transform of an image provides a frequency domain representation, where each point in the transformed image corresponds to a specific spatial frequency

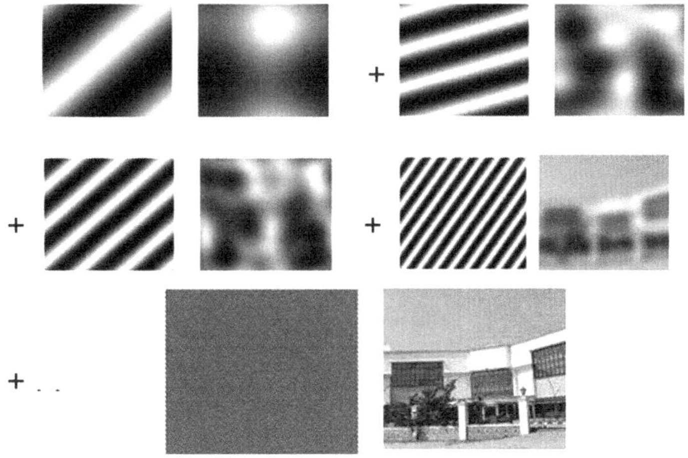

Fig. 2 A series of grayscale image pairs is arranged in a grid, each featuring distinct patterns. In each pair, the first image shows a spatial frequency pattern, while the second displays the Fourier transform of the cumulative sum of all frequency patterns up to that point. As more frequency components—ranging from low to high frequencies—are incrementally added, the Fourier image gradually takes shape. Ultimately, this process reconstructs a detailed image of the FReT-B building at TIFR Hyderabad. The arrangement visually demonstrates how combining various spatial frequencies can produce a high frequency contained image in Fourier space

component. This representation is essential for understanding how different spatial frequencies contribute to the overall image and how optical systems process these frequencies.

The frequency domain representation of an image is for understanding how different spatial frequency components contribute to its overall appearance. Low-frequency components, typically located near the center of the frequency spectrum, correspond to broad, smooth intensity variations and capture the general shape, structure, and lighting of an image. In contrast, high-frequency components, found toward the edges of the spectrum, represent fine details, textures, edges, and noise, which are vital for defining sharpness and intricate features. Figure 3 demonstrates the effect of a low-pass filter, highlighting the overall features of the image, while the high-pass filter emphasizes the image's edge responses.

This representation allows for targeted manipulation of spatial frequencies using filters to enhance or suppress specific image elements. High-pass filters emphasize high-frequency details such as edges and textures by removing low-frequency background information, improving sharpness and contrast. Conversely, low-pass filters

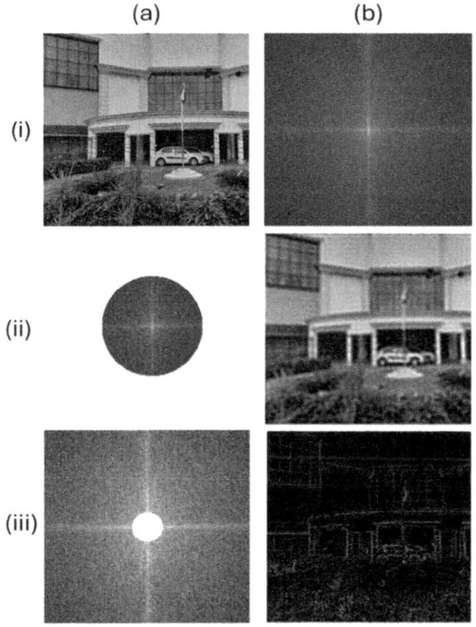

Fig. 3 Figure **i**(**a**) displays the original image of the TIFR building, while Figure **i**(**b**) shows its corresponding Fourier transform. In Figure **ii**(**a**), a low-pass frequency mask is applied to the Fourier transform in Figure **i**(**b**), selectively preserving low-frequency components while attenuating high-frequency details (such as sharp edges). The resulting filtered image, obtained after inverse Fourier transform, is shown in Figure **ii**(**b**). Similarly, Figure **iii**(**a**) illustrates the application of a high-pass frequency mask to Figure **i**(**b**), which suppresses low-frequency components and retains high-frequency details. The resulting image after inverse Fourier transform is presented in Figure **iii**(**b**)

eliminate high-frequency noise or fine details, resulting in a smoother image that retains only coarse features and broad-intensity variations [11, 12].

2.2 *Optical Transfer Function (OTF)*

The **OTF** is a fundamental metric used to quantify the ability of an optical system to reproduce spatial frequencies present in an image. It describes the system's response to various spatial frequencies and provides a comprehensive measure of its performance, particularly in terms of resolution and contrast. The OTF is defined in the frequency domain and captures how well the system transmits both the amplitude and phase of sinusoidal components corresponding to different spatial frequencies [9–14].

Mathematically, the OTF is represented as a complex function:

$$OTF(u, v) = MTF(u, v)e^{iPTF(u,v)}$$

where **modulation transfer function**, $MTF(u, v)$ describes in reduction in contrast for a given spatial frequency. It is basically the magnitude of the OTF such that $MTF(u, v) = |OTF(u, v)|$ and it takes values between 0 and 1, where 1 represents no loss in contrast. **Phase transfer function**, $PTF(u, v)$, on the other hand represents the phase shift introduced by the system for each spatial frequency. It is the phase angle of the OTF, such that $PTF(u, v) = \arg(OTF(u, v))$.

The OTF integrates both amplitude (contrast) and phase transfer, serving as a comprehensive measure of an optical system's imaging quality. High values of the Modulation Transfer Function (MTF) across spatial frequencies reflect the system's ability to resolve fine details with strong contrast, ensuring clarity and sharpness in the captured image. Conversely, notable deviations in the Phase Transfer Function (PTF) indicate phase distortions that can misalign or distort the spatial arrangement of features, undermining the fidelity of the reconstructed image. By assessing both contrast and phase accuracy, the OTF provides critical insights into system performance and its capacity to reproduce detailed and accurate visual information.

3 Spatial Frequency and Image Resolution

Image resolution is a critical measure of an optical system's ability to distinguish closely spaced details, directly tied to the spatial frequency content of an image. High spatial frequencies represent fine details, edges, and textures, while low spatial frequencies correspond to broader, smoother variations. The presence and clarity of high-frequency components define the sharpness and detail of an image. Resolution is often evaluated using the Modulation Transfer Function (MTF), which quantifies contrast preservation across spatial frequencies; the resolution limit occurs when contrast drops below a certain threshold. Factors such as diffraction, aperture size, and aberrations impose physical limits on resolution, as described by the Rayleigh criterion, where the smallest resolvable distance is inversely proportional to the numerical aperture. Digital resolution further depends on detector pixel density and adherence to the Nyquist criterion, which prevents aliasing by requiring sampling at twice the highest spatial frequency. Together, these optical and digital elements determine the system's ability to capture detailed and accurate images.

3.1 The Diffraction Limit

One of the fundamental constraints on resolution in optical imaging is the diffraction limit, a consequence of the wave nature of light. When light passes through a finite aperture, such as a lens or an optical system, it diffracts, causing point objects to spread into patterns called Airy disks. The Rayleigh criterion defines the minimum resolvable distance δ between two point sources as $\delta = \frac{1.22\lambda}{NA}$, where λ is the wavelength of light and *NA* is the numerical aperture of the imaging system. This relationship indicates that shorter wavelengths and larger numerical apertures improve resolution. The diffraction limit sets a threshold on the spatial frequencies that the system can faithfully capture, referred to as the cutoff frequency, which is given by $f_c = \frac{\mathrm{NA}}{\lambda}$. Spatial frequencies higher than this cutoff are attenuated or completely lost, leading to the inability to resolve finer details beyond the diffraction limit. As a result, the diffraction limit imposes a fundamental boundary on the sharpness and clarity of images, particularly for applications requiring the visualization of nanoscale structures or intricate details.

3.2 Super-Resolution Imaging Techniques

In recent years, significant advances in super-resolution microscopy have pushed the boundaries of optical imaging, allowing researchers to capture finer structural details at the nanoscale. Traditional optical microscopy is limited by the diffraction limit, which restricts the resolution to approximately half the wavelength of light. However, super-resolution techniques such as Structured Illumination Microscopy (SIM), Stimulated Emission Depletion (STED), and Photoactivated Localization Microscopy (PALM) have been developed to overcome this limitation. These techniques utilize sophisticated interactions between light and matter to bypass the diffraction limit, enabling the visualization of spatial frequencies beyond the traditional cutoff. By manipulating light in novel ways, these methods allow for the detection of features at scales that were previously inaccessible with conventional optical imaging.

SIM enhances resolution by using patterned light to create interference patterns that reveal finer details in the sample. STED microscopy works by selectively deactivating fluorescence in areas outside the region of interest, thus narrowing the point spread function and increasing resolution. Meanwhile, PALM exploits the precise localization of single molecules by activating and tracking them in a controlled manner, effectively reconstructing images at resolutions far beyond the diffraction limit.

SIM enhances optical resolution by projecting patterned light, such as sinusoidal grids, onto a sample and capturing multiple images with varying pattern orientations and phases. The interaction between the structured illumination and the sample's features creates Moiré fringes, which encode high spatial frequency information into lower, detectable frequencies. Computational algorithms then process these images to extract and reconstruct the high-frequency details, effectively extending the resolution beyond the diffraction limit. By utilizing this approach, SIM achieves approximately twice the resolution of conventional widefield microscopy.

4 Spatial Frequency and Image Contrast

Contrast, which refers to the difference in intensity between adjacent regions of an image, plays a vital role in determining the visibility of fine details within a sample. This contrast is closely linked to spatial frequency, as the ability to distinguish between variations in intensity relies on capturing high spatial frequencies, which correspond to finer structural details. The contrast transfer function (CTF), derived from the modulation transfer function (MTF), quantifies how contrast is preserved or altered as a function of spatial frequency. In practical imaging systems, high spatial frequencies often experience a reduction in contrast due to several factors, including optical aberrations such as chromatic and spherical distortions, diffraction limits inherent to the system, and external noise sources like thermal fluctuations or detector sensitivity. As a result, fine details encoded in high-frequency components can become blurred or invisible, leading to a loss of image sharpness and overall resolution. CTF serves as an important tool to understand and correct these effects, aiming to optimize contrast across different spatial frequencies and improve the quality of the final image.

4.1 Contrast Reduction Mechanisms

Several factors contribute to contrast reduction at high spatial frequencies in optical imaging systems. Aberrations, such as spherical aberration, astigmatism, and coma, introduce distortions that disproportionately affect higher spatial frequencies. Diffraction, as mentioned earlier, also plays a role in reducing contrast at higher frequencies by spreading light and blurring fine details. Additionally, noise, often more prominent at higher spatial frequencies, further degrades contrast.

4.2 Enhancing Contrast

Improving contrast at higher spatial frequencies is essential for enhancing image quality, particularly in applications where detecting fine details is crucial. Techniques such as apodization, which modifies the amplitude of the light passing through the optical system, can help mitigate the effects of diffraction and aberrations, thereby preserving contrast at higher spatial frequencies. Image processing methods, such as deconvolution and contrast enhancement algorithms, can also be employed to restore lost contrast during imaging.

5 Applications and Implications

The relationship between optical imaging and spatial frequency has profound implications across various fields. Understanding and controlling spatial frequency content is essential for optimizing imaging systems, developing new technologies, and improving image analysis techniques.

5.1 Medical Imaging

In medical imaging, resolving fine details and maintaining high contrast is critical for accurate diagnosis and treatment planning. Techniques such as Magnetic Resonance Imaging (MRI) and Computed Tomography (CT) rely on the spatial frequency content of images to produce high-resolution, high-contrast images of internal structures. Advanced optical imaging techniques, such as Optical Coherence Tomography (OCT), Structured Illumination Microscopy (SIM) also exploit spatial frequency principles to provide detailed images of biological tissues.

5.2 Material Science

In material science, optical imaging techniques are used to examine the microstructure of materials, detect defects, and analyze surface properties. The spatial frequency content of images plays a crucial role in resolving fine features and ensuring accurate characterization. Techniques such as Scanning Electron Microscopy (SEM) and

Atomic Force Microscopy (AFM) also rely on spatial frequency analysis to provide high-resolution images of material surfaces.

5.3 Astronomy

Astronomical imaging requires capturing images of distant objects with extremely high resolution and contrast. The spatial frequency content of these images is critical for resolving fine details, such as the structure of galaxies or the surface features of planets. Techniques such as adaptive optics, which compensates for atmospheric distortions, and interferometry, which combines signals from multiple telescopes, are used to enhance spatial frequency content and improve image quality.

5.4 Computational Imaging

Computational imaging techniques, such as digital holography and light-field imaging, leverage computational algorithms to manipulate and analyze spatial frequency content. These methods allow for the reconstruction of high-resolution images from limited data, enhancing the capabilities of traditional optical imaging systems.

6 Conclusion

The relationship between optical imaging and spatial frequency is fundamental to the design, optimization, and application of imaging systems. By understanding how spatial frequency influences resolution, contrast, and image quality, scientists and engineers can develop more effective imaging technologies and improve the analysis of captured images. As advancements in optical imaging continue to evolve, the role of spatial frequency will remain a critical factor in pushing the boundaries of what can be visualized and analyzed using light.

References

1. Choudhury, P., Goutam Buddha, S.S., Kumar, N.: Peering beyond the diffraction frontier: a guide to super-resolution microscopy. In: Materials for Electronic, Magnetic, and Spintronic Technologies: Characterization and Applications from Energy Storage to Disease Detection, pp. 171–187. Springer Nature Switzerland, Cham (2024)

2. Goutam Buddha, S.S., et al.: Advancements in confocal and array detection microscopy. In: Materials for Electronic, Magnetic, and Spintronic Technologies: Characterization and Applications from Energy Storage to Disease Detection, pp. 209–221. Springer Nature Switzerland, Cham (2024)
3. Goutam Buddha, S.S., Boruah, B.R.: Estimation of point spread function of an optical microscope using stochastic minimization of least square errors. J. Opt. **22**(5), 055603 (2020)
4. Goutam Buddha, S.S.: Resolution enhancement in a scanning optical microscope using array detection (2020)
5. Goutam Buddha, S.S., Kalita, R., Boruah, B.R.: Estimation of point spread function of an imaging system using a programmable target. In: Three-Dimensional and Multidimensional Microscopy: Image Acquisition and Processing XXV, vol. 10499. SPIE (2018)
6. Goutam Buddha, S.S., et al.: Subtraction microscopy using aberrated PSFs for image contrast enhancement. In: 2023 IEEE Workshop on Recent Advances in Photonics (WRAP). IEEE (2023)
7. Goutam Buddha, S.S., et al.: Estimation of 3D point spread function of a widefield microscope by stochastic least square minimization method. In: Three-Dimensional and Multidimensional Microscopy: Image Acquisition and Processing XXX, vol. 12385. SPIE (2023)
8. Goodman, J.W.: Introduction to Fourier Optics, 3rd edn. Roberts & Company Publishers (2005)
9. Born, M., Wolf, E.: Principles of Optics, 7th edn. Cambridge University Press (1999)
10. Hecht, E.: Optics, 5th edn. Pearson (2016)
11. Abbe, E.: Beiträge zur Theorie des Mikroskops und der mikroskopischen Wahrnehmung. Arch. Mikrosk. Anat. **9**(1), 413–418 (1873)
12. Schermelleh, L., Heintzmann, R., Leonhardt, H.: A guide to super-resolution fluorescence microscopy. J. Cell Biol. **190**(2), 165–175 (2010)
13. Booth, M.J.: Adaptive optics in microscopy. Philos. Trans. R. Soc. A Math. Phys. Eng. Sci. **365**(1861), 2829–2843 (2007)
14. Gabor, D.: A new microscopic principle. Nature **161**(4098), 777–778 (1946)

Fundamentals of Ray Tracing and Lens Design With Zemax OpticStudio

Karuna Sindhu Malik, Shilpa Singh, Nagendra Kumar, SSGoutam Buddha, and Santanu Konwar

Abstract Ray tracing is a fundamental technique in optical system analysis and design, enabling the modeling of light propagation through lenses, mirrors, and other optical elements. Paraxial ray tracing operates under the small-angle approximation, which assumes that light rays make small angles with the optical axis, simplifying calculations by ignoring higher-order terms. This method is highly effective for preliminary optical system design, offering quick insights into image formation and magnification. However, its limitations become evident when dealing with systems requiring high precision where aberrations and distortions are significant. Exact ray tracing, on the other hand, accounts for the full geometry of light rays without approximations, providing a more accurate representation of real-world optical behavior. It considers all ray intersections and refractions according to Snell's Law, making it important for designing complex systems such as objective lens, telescopes, microscopes, and cameras. In this chapter we will study the paraxial and exact ray tracing through optical system. We will design few lenses using softwares like Python and Zemax OpticStudio.

Keywords Ray tracing · Optical axis · Microscopes · Optical devices

K. S. Malik (✉)
Department of Physics, Brainware University, Barasat, Kolkata-700125, West Bengal, India
e-mail: karunasindhu7@gmail.com; ksm.ph@brainwareuniversity.ac.in

S. Singh
Department of Physics, Indian Institute of Technology Guwahati, Assam, India

N. Kumar
Department of Physics, Indian Institute of Technology Delhi, New Delhi, India

S. Buddha
Department of Physics, Tata Institute of Fundamental Research Hyderabad, Hyderabad, India

S. Konwar
Department of Physics, Abhayapuri College, Abhayapuri, Bongaigaon, Assam, India

A. Mishra and V. Pathak (eds.), *Smart Materials for Energy Storage and Biomedical Applications*, Engineering Materials, https://doi.org/10.1007/978-3-031-92584-9_12

1 Introduction

Ray tracing (RT) is a fundamental method used to model and understand the behavior of light as it passes through optical systems [1–4]. It is an essential tool in the design and analysis of various optical devices such as lenses, mirrors, telescopes, microscopes, and other imaging systems. Ray tracing is important to calculate optical forces in optical trapping (OT) [1, 5], detection and correction of aberration in wavefront sensing [6–8] and study of image quality, magnification in microscopy [9, 10] etc. The technique has developed significantly over time. Around 17th to eighteenth century, scientists like Kepler and Snell have developed basic foundation of geometrical optics [11, 12]. For simpler understanding, paraxial approximations were used that allowed for basic ray tracing in simple optical system. Around the beginning of twentieth century, scientists like Carl Zeiss and Ernst Abbe used analytical ray tracing that helped in improving optical design [13]. A breakthrough in ray tracing occurred with the advent of computers in mid-twentieth century, where the rays were numerically traced efficiently for complex optical system. Over the following years, ray tracing became more advanced with the optimizing ray tracing software such as Zemax, OSLO etc. opening door to different applications in various scientific fields [14–17].

In paraxial ray tracing (PRT), rays traveling at small angles relative to the optical axis are accurately traced using matrix formalism, such as the Jones matrix approach [18]. Numerous studies on paraxial approximations have demonstrated their effectiveness for simpler optical systems, where small-angle approximations hold true [19–21]. However, for practical and more complex implementations, exact ray tracing (ERT) methods are employed. These methods track the trajectories of skew rays without relying on approximations, providing a more precise description of light propagation [22–26]. Over time, extensive scientific research has been conducted to refine and expand these exact ray tracing techniques.

The mathematical framework of ray tracing has further evolved to incorporate the polarization properties of light, thus, enabling researchers to study the behavior of light beams in various imaging scenarios [27–29]. This has facilitated advancements in the fabrication and optimization of intricate optical systems. Recent studies have leveraged these advancements to achieve higher precision in designing and analyzing complex optical components.

Additionally, these exact ray tracing principles have been incorporated into powerful optical design software, such as Zemax and OSLO [30–32]. These softwares simulate a vast number of rays and apply advanced optimization algorithms to deliver accurate and detailed image analyses. By modeling the trajectories of light across an optical system, these software solutions have significantly improved the design and evaluation of modern optical devices, making them indispensable for the field of optical engineering.

2 Theory of Ray Tracing Through Optical System

In this section, we will study the theoretical foundation to gain a deep understanding of the concept of geometrical ray tracing. A detailed explanation of the basic principles, for both paraxial ray tracing and exact ray tracing, is provided below. We will also discuss about the sequential and non-sequential ray tracing through Zemax OpticStudio.

2.1 ABCD Matrix Method of Paraxial Ray Tracing

Paraxial ray tracing is a simplified approach used in geometrical optics to model the behavior of light rays close to the optical axis. It assumes that the light rays make small angles with the optical axis, so their paths can be approximated by straight lines and their angles are considered to be small enough to ignore higher-order terms in the ray's equations. Therefore, while considering refraction through a spherical lens we need to impose certain restriction so that the above condition is satisfied for every set of rays. That is, we use an approximation where,

$$\sin\theta \sim \theta \tag{1}$$

where θ is the angle between a ray and optical axis of the optical system. This approximation is known as paraxial approximation. This approximation is used to arrive at the lens maker formula [33] which is given as

$$\frac{1}{f} = \frac{1}{v} - \frac{1}{u} = \left(\frac{\mu_2}{\mu_1} - 1\right)\left(\frac{1}{R_1} - \frac{1}{R_2}\right) \tag{2}$$

where f is focal length of the lens, μ_1 and μ_2 are refractive indices of surrounding medium and lens material respectively,

R_1 and R_2 are radius of curvature of the first and second surface of lens and u and v are object and image distances respectively.

Paraxial approximation makes the calculation involved in geometrical optics quite simpler and also helps in getting rid of optical aberration [4]. The ABCD ray transfer matrix is a mathematical tool used in optical systems to describe the propagation of light rays through various optical elements, such as lenses, mirrors, and free space. It is part of a broader approach called geometrical optics, which treats light as rays rather than waves. They are used to trace rays travelling only in one plane containing the optical axis (meridional rays) as well as the rays travelling in more than one plane. To define meridional rays, we just need two parameters.

1. The height of the ray with respect to the optical axis (y).
2. The angle made by the ray with respect to the optical axis (θ).

For a light ray with position y and angle θ the matrix equation is

$$\begin{pmatrix} y_2 \\ \theta_2 \end{pmatrix} = \begin{pmatrix} A & B \\ C & D \end{pmatrix} \begin{pmatrix} y_1 \\ \theta_1 \end{pmatrix} = M \begin{pmatrix} y_1 \\ \theta_1 \end{pmatrix} \quad (3)$$

where y_1 and θ_1 are the initial position and angle, y_2 and θ_2 are the final position and angle, $M = \begin{pmatrix} A & B \\ C & D \end{pmatrix}$ is the ray transfer matrix or ABCD matrix (Fig. 1) (Table 1).

In the above discussion we have seen the 2×2 ray transfer matrix for various optical propagation. Sometimes we use 3×3 ray transfer matrix for non-reflecting system in homogeneous coordinate system. In 3×3 ray transfer matrix the extra parameters useful for understanding of rotated and decentered optical elements (Table 2).

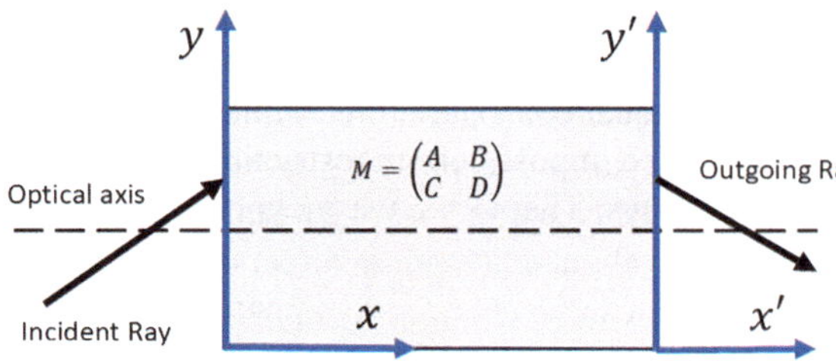

Fig. 1 The incident ray transforms into outgoing ray through the gray box by ABCD ray transfer matrix. 2 × 2 ABCD matrices of few optical elements are tabulated below

Table 1 ABCD matrices of some optical elements

ABCD matrix	Description
$M = \begin{pmatrix} 1 & \frac{d}{\mu} \\ 0 & 1 \end{pmatrix}$	Ray translation through homogeneous medium. Ray travels d distance through medium of refractive index μ.
$M = \begin{pmatrix} 1 & 0 \\ 0 & \frac{\mu_1}{\mu_2} \end{pmatrix}$	Refraction at plane surface or boundary. μ_1 and μ_2 are refractive indices of first and second medium respectively.
$M = \begin{pmatrix} 1 & 0 \\ \frac{\mu_1-\mu_2}{\mu_2 R} & \frac{\mu_1}{\mu_2} \end{pmatrix}$	Refraction at spherical surface or boundary. μ_1 and μ_2 are refractive indices of first and second medium respectively and R is the radius of curvature of spherical surface.
$M = \begin{pmatrix} 1 & 0 \\ -\frac{1}{f} & 1 \end{pmatrix}$	Refraction through thin lens. f is focal length of the thin lens.
$M = \begin{pmatrix} 1 & 0 \\ 0 & 1 \end{pmatrix}$	Reflection from a plane mirror.
$M = \begin{pmatrix} 1 & 0 \\ \frac{2}{R} & 1 \end{pmatrix}$	Reflection from a spherical mirror. R is radius of curvature of spherical mirror.

Table 2 ABCD (3 × 3) matrices

ABCD matrix	Description
$M = \begin{pmatrix} 1 & 0 & 0 \\ 0 & \cos\alpha & -\sin\alpha \\ 0 & \sin\alpha & \cos\alpha \end{pmatrix}$	Rotation by angle α
$M = \begin{pmatrix} 1 & -d_1 & -d_2 \\ 0 & 1 & 0 \\ 0 & 0 & 1 \end{pmatrix}$	Displacement (d_1, d_2) matrix

2.2 Exact Ray Tracing

Exact ray tracing is a theoretical concept based on equations that allow us to compute the precise path of skew rays without any approximation [34, 35]. These equations can ultimately be implemented numerically to provide a detailed representation of the optical systems.

To trace a ray using the exact ray tracing technique, we just need two parameters.

1. The initial coordinate of incident ray $P = (x_i, y_i, z_i)$ and
2. The initial direction cosines of incident ray (l_i, m_i, n_i) which is shown in Fig. 2.

Figure 2 shows that a ray with starting coordinate $P = (x_i, y_i, z_i)$ and direction cosines (l_i, m_i, n_i) intersect the spherical surface at a point $Q = (x_f, y_f, z_f)$. Here, n_1 and n_2 are the refractive indices before and after the spherical surface. R is radius of curvature, θ_i and θ_r are the incident and refracted angle of the ray.

If the distance $PQ = \Delta$ then it can be written that

$$x_f = x_i + \Delta l_i$$

$$y_f = y_i + \Delta m_i$$

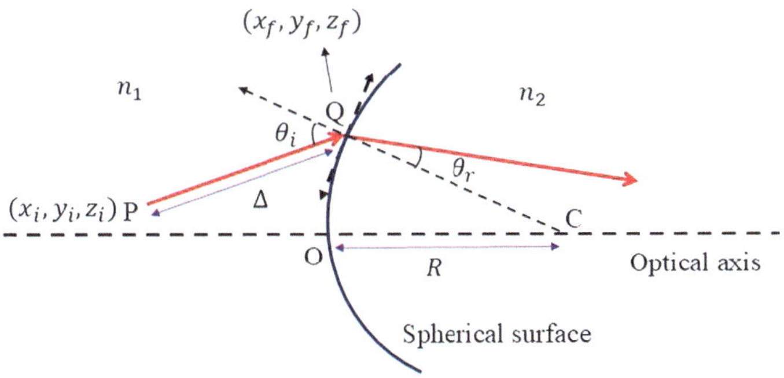

Fig. 2 Schematic diagram of exact ray tracing through a spherical surface of radius of curvature R

$$z_f = z_i + \Delta n_i \tag{4}$$

The final direction cosines of the refracted ray in the second medium is

$$\begin{aligned} l_f &= sl_i + \alpha_{ext}(\cos\theta_r - s\cos\theta_i) \\ m_f &= sm_i + \beta_{ext}(\cos\theta_r - s\cos\theta_i) \\ n_f &= sn_i + \gamma_{ext}(\cos\theta_r - s\cos\theta_i) \end{aligned} \tag{5}$$

where $s = \frac{n_1}{n_2}$, $\alpha_{ext} = -cx_i$, $\beta_{ext} = -cy_i$, $\gamma_{ext} = (1 - cz_i)$.

Equations 4 and 5 are the translation and refraction equations for the exact ray tracing. These equations are important to calculate all the intersection coordinates on the lens surface one after another using exact ray tracing if we know the lens parameters like radii of curvature, center thickness, back focal length and also the refractive index of the medium.

2.3 Ray Tracing Through Zemax OpticStudio

Ray tracing in Zemax OpticStudio [36] is a powerful method used to model the propagation of light through optical systems, simulating how rays travel and interact with surfaces, such as lenses or mirrors. Zemax OpticStudio utilizes ray tracing to analyze optical performance, identify aberrations, and optimize designs. Two common types of ray tracing in Zemax are sequential ray tracing and non-sequential ray tracing. Sequential ray tracing follows rays through a system step-by-step, from one surface to the next in a predefined sequence. However, in non-sequential ray tracing sources, objects, and detectors are used in 3D space, with no predefined order for the rays to travel through the system.

Figure 3a demonstrates the sequential ray tracing process. Ray 1, emitted from the laser, strikes a beam splitter (BS), and the reflected ray (ray 2) then reaches a detector. Each ray interacts with the optical elements in a sequential manner. In contrast, Fig. 3b depicts the non-sequential ray tracing process. Ray 1, emitted from the laser, strikes the beam splitter (BS), and the reflected ray (ray 2) hits a mirror. The mirror reflects ray 4, which then passes through the BS again, splitting into two components: the reflected ray (ray 5) and the transmitted ray (ray 6). This process, where rays intersect the BS twice, illustrates the non-sequential nature of the ray tracing.

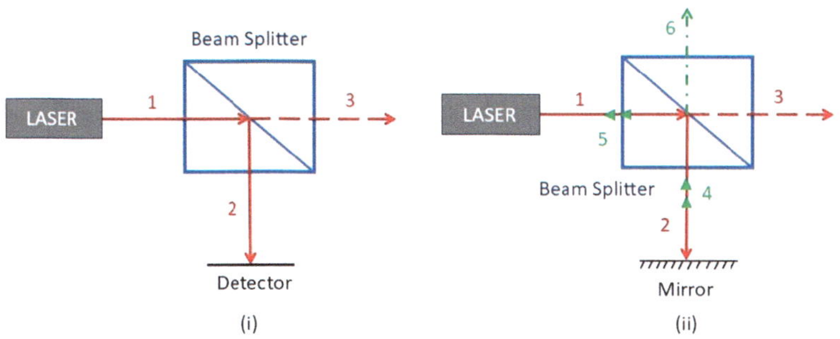

Fig. 3 Schematic diagram of **(i)** sequential and **(ii)** non-sequential ray tracing

3 Simulation Result of Ray Tracing Through Optical System

We have performed the Python based numerical simulation of paraxial ray tracing and exact ray tracing. We have presented the simulation results by the following three different representations such as ray diagram, spot diagram and the corresponding intensity plot (line-plot) which are shown in Figs. 4 and 5. For ray tracing we have considered a plane convex lens, from Thorlabs [37], whose diameter is 25.4 mm and back focal length is 59.88 mm. For performing the paraxial ray tracing we have used the ABCD translation matrix, refraction matrix at plane and spherical boundary. Figure 4a shows the paraxial ray tracing through plano convex lens where all the rays are converging at a fixed point (shown inside red box). Figure 4b (i–iii) show the spot diagram of ray tracing when the image plane is 4 mm before the actual focal plane, at the nominal focal plane and 4 mm after the nominal focal plane. Figure 4c (i–iii) show the intensity plot or line plot corresponding to each spot diagram. The intensity plots are consistent with the expected results which depict the variation of intensity gradient at the focal volume corresponding to each spot diagram. We also performed the geometrical exact ray tracing of the same plano convex lens. Figure 5 shows the exact ray tracing results where we can see that marginal rays and paraxial rays are converging at different points. The behavior of rays is due to spherical aberration (shown inside the red box). Figure 5b (i–iii) display the spot diagram of ray tracing for three different positions of the image plane-4 mm before the nominal focal plane, at the nominal focal plane, and 4 mm after the nominal focal plane. Figure 5c (i–iii) present the corresponding intensity or line plots for each spot diagram at the focal volume.

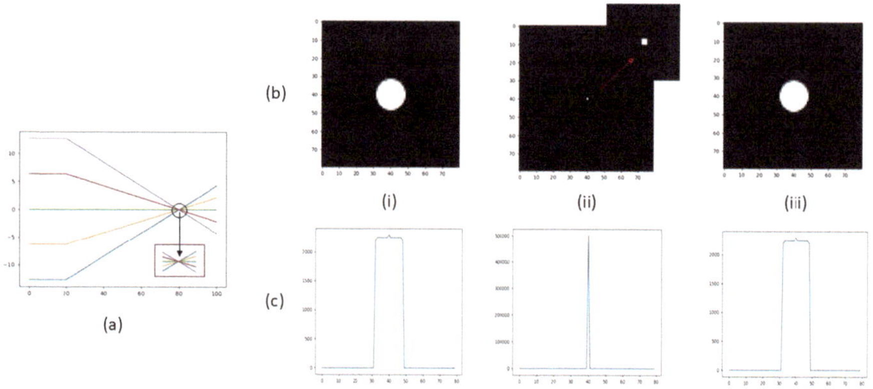

Fig. 4 **a** represents paraxial ray tracing through a plano convex lens. **b** (i–iii) illustrate the spot diagram when the image plane is placed at 4 mm before the focal plane, at nominal focal plane and 4 mm after the focal plane. **c** (i–iii) show the intensity plot corresponding to each spot diagram

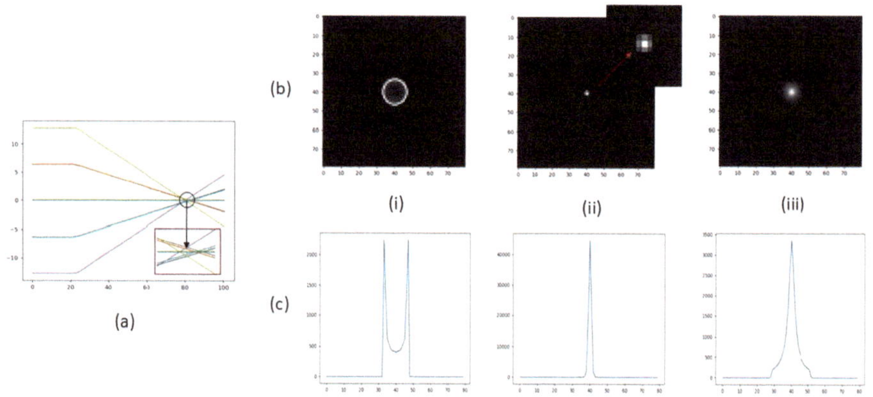

Fig. 5 **a** represents exact ray tracing through a plano convex lens. **b** (i–iii) illustrate the spot diagram when the image plane is placed at 4 mm before the focal plane, at nominal focal plane and 4 mm after the focal plane. **c** (i–iii) show the intensity plot corresponding to each spot diagram

4 Lens Design by Zemax OpticStudio

To understand the ray tracing preciously we also performed the ray tracing in Zemax OpticStudio. We have designed a telephoto lens [38] of effective focal length 500 mm (approx.). A telephoto lens is a type of camera lens with a long focal length, typically ranging from 70 mm to 300 mm or more, that allows photographers to capture distant subjects with greater magnification. Telephoto lenses are commonly used in wildlife, sports, and portrait photography, where subjects are far away and a close-up view is desired [39]. We have considered two lens (from Thorlabs) one is plano convex lens

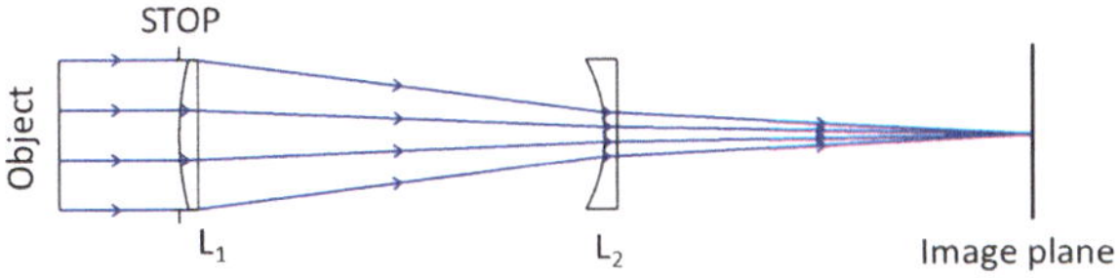

Fig. 6 Design of telephoto lens by Zemax OpticStudio

(radius = 103 mm, thickness = 6.2 mm, N-BK7 glass) and another is plano concave lens (radius = − 51.5 mm, thickness = 4 mm, N-BK7 glass).

Figure 6 shows the design of a telephoto lens of effective focal length (EFL) = 499.95 mm. Here the diameter of the aperture stop is 50 mm, the distance between L_1 (plano convex lens) and L_2 (plano concave lens) is 134.7 mm.

5 Conclusion

This chapter provides a comprehensive overview of ray tracing through optical system particularly in lens system. We have discussed two types of ray tracing one is paraxial and another is exact ray tracing. The discussion of paraxial ray tracing is based on ABCD translation matrix. The ABCD translation matrix helps to understand the ray tracing when the rays make small angle with the optical axis. However, the geometrical exact ray tracing is more effective ray tracing as it traces both paraxial and marginal rays. We have presented few Python based simulation results of paraxial and exact ray tracing of a plano convex lens. Zemax OpticStudio is one of the powerful ray tracing software is available in market. So, we also design a telephoto camera lens of effective focal length = 499.95 mm. All the lens information are taken from Thorlabs. Overall, this chapter highlights the significance of ray tracing technique through optical system.

References

1. Sindhu Malik, K., Boruah, B.R.: Optical force calculation in the ray-optics regime for beams with arbitrary complex amplitude profiles. Opt. Lett. **47**(16), 4151–4154 (2022)
2. Malik, K.S., Konwar, S., Goutam Buddha, S.S., Kumar, N.: Computation of spot diagrams and ray tracing of different aberrated beams using geometrical exact ray tracing. In 2023 IEEE Workshop on Recent Advances in Photonics (WRAP), pp. 1–3. IEEE, (2023)
3. Malik, K.S., Goutam Buddha, S.S., Kumar, N.: Calculation of optical forces in optical tweezers. In Materials for Electronic, Magnetic, and Spintronic Technologies: Characterization and Applications from Energy Storage to Disease Detection, pp. 223–235. Cham: Springer Nature Switzerland, (2024)
4. Malik, K.S., Boruah, B.R.: Calculation of spot diagrams of different vector beams using geometrical ray tracing. In Complex Light and Optical Forces XIV, vol. 11297, pp. 30–36. SPIE, (2020)
5. Malik, K.S., Boruah, B.R.: Optimal trap velocity in a dynamic holographic optical trap using a nematic liquid crystal spatial light modulator. J. Opt. **24**(3), 034004 (2022)

6. Kumar, N., Choudhury, P., Sherpa, N., Chauhan, A., Goutam Buddha, S.S., Malik, K.S., Konwar, S., Boruah, B.R.: Improvement in wavefront measurement accuracy in grating array-based zonal wavefront sensors by modulating the intensity profile of the incident beam. In Practical Holography XXXVII: Displays, Materials, and Applications, vol. 12445, pp. 173–180. SPIE, (2023)
7. Kumar, N., Pathak, B., Kesarwani, R., Goswami, S., Khare, A., Boruah, B.R.: Experimental demonstration of in situ surface and thickness profile measurements of thin film during deposition using a grating array based wavefront sensor. Opt. Lett. **47**(21), 5509–5512 (2022)
8. Kumar, N., Khare, A., Boruah, B.R.: Enhancement of wavefront measurement sensitivity in a zonal wavefront sensor without curtailing the sensing speed. J. Opt. **24**(1), 014003 (2021)
9. Buddha, S.G., Kumar, N., Malik, K.S., Konwar, S.: Subtraction microscopy using aberrated PSFs for image contrast enhancement. In 2023 IEEE Workshop on Recent Advances in Photonics (WRAP), pp. 1–3. IEEE, (2023)
10. Buddha, S.G., Kumar, N., Choudhury, P., Malik, K.S.: Advancements in confocal and array detection microscopy. In Materials for Electronic, Magnetic, and Spintronic Technologies: Characterization and Applications from Energy Storage to Disease Detection, pp. 209–221. Cham: Springer Nature Switzerland, (2024)
11. Vollgraff, J.A.: Snellius notes on the reflection and refraction of rays. Osiris, vol. 1, pp. 718–25. JSTOR, (1936). http://www.jstor.org/stable/301634. Accessed 6 Dec. 2024
12. Swinden, B.A.: 2391 Johann Kepler: Paralipomena Ad Vitellionem. Math. Gaz. **38**(323), 44–46 (1954). https://doi.org/10.2307/3609779
13. Feffer, S.M.: Microscopes to munitions: Ernst Abbe, Carl Zeiss, and the transformation of technical optics, 1850–1914. University of California, Berkeley (1994)
14. Krokstad, A., Strom, S., Sørsdal, S.: Calculating the acoustical room response by the use of a ray tracing technique. J. Sound Vib. **8**(1), 118–125 (1968)
15. Gjøystdal, H., et al.: Review of ray theory applications in modelling and imaging of seismic data. Stud. Geophys. et Geod. **46**, 113–164 (2002)
16. Virieux, J., Farra, V.: Ray tracing in 3-D complex isotropic media: an analysis of the problem. Geophysics **56**(12), 2057–2069 (1991)
17. Yun, Z., Iskander, M.F.: Ray tracing for radio propagation modeling: principles and applications. IEEE Access **3**, 1089–1100 (2015)
18. Goodman, J.W.: Introduction to Fourier optics. Roberts and Company Publishers, (2005)
19. Tanaka, K.: II Paraxial theory in optical design in terms of Gaussian brackets. Prog. Opt. **23**, 63–111 (1986)
20. Preussner, P.R., et al.: Ray tracing for intraocular lens calculation. J Cataract. Refract. Surg. **28**(8), 1412–1419 (2002)
21. Fujiwara, E., Cordeiro, C.M.B.: Model-based design and simulation of paraxial ray optics systems. Appl. Sci. **10**(22), 8278 (2020)
22. Richards, B., Wolf, E.: Electromagnetic diffraction in optical systems ii. Structure of the image field in an aplanatic system, Proceedings of the Royal Society of London. Series A Math. Phys. Sci. 1274, 358–379 (1959)
23. Kidger, M.J.: Fundamental optical design, SPIE Bellingham, (2001)
24. Lin, P.D., Lu, C.H.: Analysis and design of optical systems by use of sensitivity analysis of skew ray tracing. Appl. Opt. **43**(4), 796–807 (2004)
25. Einighammer, J., et al.: Calculating intraocular lens geometry by real ray tracing. J. Refract. Surg. **23**(4), 393–404 (2007)
26. Conrady, A.E.: Applied optics and optical design, part one. Courier Corporation, (2013)
27. Chipman, R.A.: Mechanics of polarization ray tracing. Opt. Eng. **34**(6), 1636–1645 (1995)
28. Foreman, M.R., P. T¨or¨ok, Computational methods in vectorial imaging, J. Mod. Opt. 58, 5–6, 339–364, (2011)
29. Kim, J., Wang, Y., Zhang, X.: Calculation of vectorial diffraction in optical systems. JOSA A **35**(4), 526–535 (2018)
30. Pineau No¨el, V., Masoumi, S., Parham, E., Genest, G., B´egin, L., Vigneault, M.A., C^ot´e, D.C.: Tools and tutorial on practical ray tracing for microscopy, Neurophotonics, 8, 1, 010801–010801, (2021)

31. Sarbazi, E., et al.: Ray tracing based channel modeling for visible light communications. 2014 22nd Signal Processing and Communications Applications Conference (SIU). IEEE, (2014)
32. Bösel, C., Gross, H.: Ray mapping approach for the efficient design of continuous freeform surfaces. Opt. Express **24**(13), 14271–14282 (2016)
33. Pedrotti, F.L., Pedrotti, L.M., Pedrotti. L.S.: Introduction to optics. Cambridge University Press, (2018)
34. Malik, K.S.: Augmentation of ray pencil model to calculate optical trapping force for arbitrary beam profiles and its experimental validation. (2022)
35. Welford, W.T.: Aberrations of optical systems. Routledge, (2017)
36. Laikin, M.: Lens design. Crc Press, (2018)
37. https://www.thorlabs.com/newgrouppage9.cfm?objectgroup_id=112
38. Kingslake, R.: Telephoto versus ordinary lenses: a tutorial paper. J. SMPTE **75**(12), 1165–1168 (1966)
39. Ray, S.: Applied photographic optics. Routledge, (2002)

Metal Oxide Nanostructures for Optoelectronic Applications

Sakshi Bagwari, Manish Jha, Sachin Pathak, Dheeraj K. Singh, Amodini Mishra, and Vishakha Kaushik

Abstract The fields of optoelectronics have opened up new ways to make technologies that are very important in today's world. From the atomic level to the development of an optical system, research has gone from being very basic to very advanced. A lot of people are interested in metal oxides (MOs) like ITO (Indium Tin Oxide), zinc oxide (ZnO), and barium oxide (BaO) based nanomaterials because they have unique qualities and could be explored in optoelectronics devices like solar cells, OLEDs, flat-panel displays, and flexible transparent electronics. In terms of thin-film production, optoelectronic properties, electronic structure, defect states, charge transport methods, and materials design principles, MOs semiconductors are very distinct from inorganic semiconductors like silicon and III-V compounds. MOs are different from other unusual electronic materials. This chapter talks about how they are unique and how they are the same with respect to their physicochemical properties, electrical properties, thin-film production techniques, and their specific uses in solar cells and thin-film transistors.

Keywords Metal oxides · Electrical properties · Thin film transistors and solar cells

S. Bagwari · M. Jha · V. Kaushik (✉)
Materials and Nano Engineering Research Laboratory, Department of Physics, School of Physical Sciences, DIT University, Dehradun, India
e-mail: vishakha.kaushik@dituniversity.edu.in

S. Bagwari
Department of Physics, School of Advanced Engineering, UPES, Dehradun, Uttarakhand, India

S. Pathak
Department of Physics, School of Engineering, University of Petroleum and Energy Studies, Dehradun, India

D. K. Singh
School of Nano Sciences, Central University of Gujarat, Gandhinagar, Gujarat, India

A. Mishra
Department of Physics, Indian Institute of Technology Delhi, New Delhi, India

A. Mishra and V. Pathak (eds.), *Smart Materials for Energy Storage and Biomedical Applications*, Engineering Materials, https://doi.org/10.1007/978-3-031-92584-9_13

1 Introduction

From 1960s onwards, the features of semiconducting metal oxides led to them being utilised as gas sensors. The gas sensors based on metal oxides demonstrated high sensitivity, quick response time and also were cost effective. Metallic oxide nanoparticles (MONPs) possess various features like catalytic action, superparamagnetism, greater ductility, selectivity of nanophases, larger surface area available for reaction etc. which make them crucial for numerous nanotechnological applications. As compared to the bulk material, the nanomaterial of similar composition has lower melting point [1]. Regarding synthetic methodologies, there are generally two kinds of techniques for to preparing nanomaterials and nanostructures. Metal oxides show promising electrical and optical properties [2]. Based on electrical conductivity: Conductors, semiconductors, and insulators are the three categories into which metal oxides fall. For example-conductor; oxides like tin(II) oxide (SnO) can conduct electricity. Semiconductors; Indium tin oxide (ITO) and zinc oxide (ZnO) are mostly used in electronics. Insulators: Oxides such as alumina (Al_2O_3) and silica (SiO_2) are excellent electrical insulators. Further, metal oxides can be used in optical applications since they are transparent to visible light. ITO, for instance, is utilized in touchscreens and displays due to its conductivity and transparency. Another important property is the chemical stability. Metal oxides can withstand high temperatures and severe chemicals since they are typically stable and corrosion-resistant.

Due to their exceptional properties, they are utilized in many applications like catalysis: Metal oxides are employed in a variety of chemical processes as catalysts. For instance, titanium dioxide (TiO_2) is a catalyst used in environmental clean-up and photocatalysis. Electronics: oxides like ITO are transparent and conductive, they are utilized in solar cells, LEDs, and touchscreens. Metal oxides are used in supercapacitors and batteries for energy storage. For example, manganese dioxide (MnO_2) is a component of alkaline batteries. In the next part, we will discuss the synthesis methods and optoelectronic applications for MOs.

2 Synthesis Methods for MOs

Nanoparticle synthesis is accomplished through a wide range of techniques. It is possible to differentiate three primary subgroups: mechanical techniques like milling, physical methods, and chemical approaches. The main methods and synthesis approaches utilized in nanotechnology are outlined in Fig. 1.

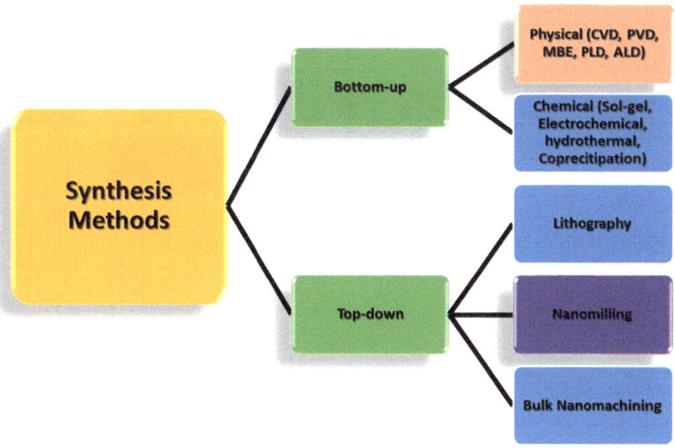

Fig. 1 Synthetic strategies showing typical processes for making MOs

2.1 Bottom-Up Approach

The bottom-up strategy for synthesizing metal oxides entails building them from smaller components, like atoms or molecules. The composition, structure, and characteristics of the metal oxides can be precisely controlled using this approach. Utilizing and combining sub- or nanoscale materials (atoms or molecules), the Bottom-up method produces nanostructures with novel or altered functionality. This approach might be considered the natural development of supramolecular chemistry, which is the functional and structural study of the substances formed by the accumulation of multiple chemical species and concentrates on intermolecular connections. It also allows for a more regulated system design. Conversely, the bottom-up approach may result in less waste and, therefore, be more economical. The bottom-up method is a process of creating of nanoparticles by accumulating atoms, molecules into larger form. The major drawback of this approach is that large scale synthesis is not possible due to complex chemistry associated multi-component structures. The bottom-up, or self-assembly, approach to nanofabrication builds larger structures from smaller ones by utilizing chemical or physical forces at the nanoscale. As component sizes decrease, the bottom-up approach in nanofabrication is becoming an increasingly important complement to top-down techniques.

An outline of the main methods utilized for bottom-up and top-down preparation of metal oxides has been provided below:

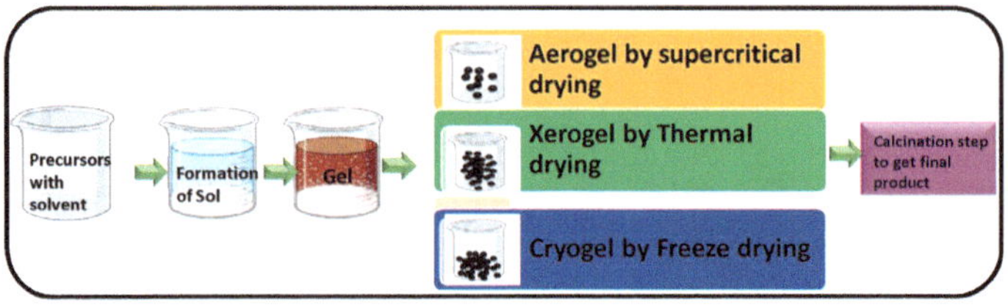

Fig. 2 Schematic of different stages of the sol–gel process

2.1.1 Sol–gel Method

This process can help in production of nanoparticles of uniform size on commercial scale. When metallic oxide precursors are combined in distinct ratios, this technique can create various kinds of nanoparticles simultaneously and ensure single step creation of alloys [3]. Other techniques, such as the electrochemical and plasma approaches could also result in one step synthesis of alloys. Compared to these methods, sol gel could be utilised for industrial scale synthesis [4]. For this reason, sol gel is arguably the most utilised method for synthesizing metal oxide nanoparticles. Here, "sol" denotes the solution after the MO dissolves in the organic solvent. The hydrolysis, condensation, and polymerization processes make up the sol–gel transition. A three-dimensional gel that is dense and porous is produced as the hydroxide polymerizes. Finally, the oxide form could be adjusted by subjecting it to further drying and heating; this could result in coatings, fibres, solid foam, nanoparticles, etc., as shown in Fig. 2.

2.1.2 Synthesis of Zinc Oxide Nanoparticle (ZnO)

Zinc oxide (ZnO) has excellent optical and electrical qualities in addition to being inexpensive, non-toxic, and abundant in nature, it is a very appealing substitute material for ITO [5]. At ambient temperature, ZnO exhibits a significant value (– 60 meV) of exciton binding energy o and a relatively wide band gap energy value (– 3.37 eV), making it an *n*-type semiconductor. As compared to ITO, it has good stability in hydrogen plasma, high thermal stability, and high electrochemical stability [6]. Prior research has also documented ZnO thin film's strong electrical stability [7], and its application as an air-stable anode for organic light-emitting diodes (OLEDs). Using a range of depositing processes, ZnO thin film nanostructures, including nanoparticles, nano-rings, micro-rings and micro-rods, have been effectively produced. Radio frequency magnetron sputtering [8], spray pyrolysis, dip coating, spin coating and electron beam evaporation are some of the deposition methods that have been studied the most. Out of them, spray pyrolysis is particularly interesting since it can coat huge surfaces at a low cost, is simple, safe, doesn't require a high vacuum, and produces less waste [9]. The molecular form of ZnO is shown in Fig. 3.

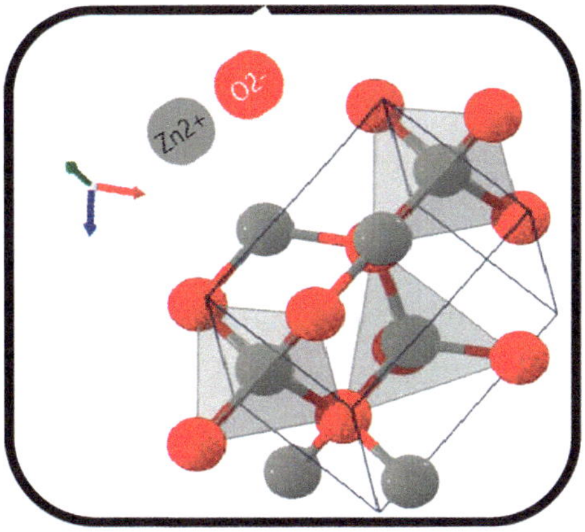

Fig. 3 Schematic shows the illustration of ZnO in molecular form

To prepare the zinc oxide, 2 g of zinc acetate $Zn(CH_3COO)_2$ and 8 g of sodium hydroxide (NaOH) were mixed with distilled water separately. The zinc acetate was placed in 10 ml of distilled water, and the sodium hydroxide was placed in 15 ml of distilled water. These solutions were stirred separately for 5 min with a constant stirring and then these solutions were again stirred for 5 min together with a constant stirring. Afterwards, a burette containing 100 ml ethanol was titrated dropwise to the final solution which contained mixture of both the solutions. After the reaction white precipitate was formed.

2.1.3 Structure and Morphology

The nature, structure, physiochemistry and morphology of the produced NPs were examined utilising numerous analytical techniques, including Scanning Electron Microscopy (SEM), and X-ray diffraction (XRD). Figure 4 shows the SEM and TEM micrographs of ZnO nanoparticles at different magnifications. SEM was utilised in examining the topography of the surface and compositions of ZnO NPs. FTIR spectroscopy, which has a range of 500–4000 cm^{-1}, proved suitable for providing details regarding the identity of the molecular component. The analysis was also conducted on the ZnO NPs dry powder. Through XRD, as shown in Fig. 5, the information regarding the size of the NPs and the phase of the crystalline material can be obtained.

Other synthesis methods include physical synthesis, where a solid precursor is broken down into molecules, individual atoms, which in turn are deposited in the desired shape of the nanostructure. But these procedures are typically the priciest since they necessitate sophisticated vacuum techniques and have a strictly chamber-dimension-limited deposit area. There is a wide range of physical deposition techniques available, and the physical vapour deposition method has been effectively used to create MONPs: In physical vapour synthesis, a solid target is exposed to a plasma arc, which produces vapour and sets off reactions that result in supersaturation. Upon cooling down and combining the resulting atoms with the reactive gas,

Fig. 4 Low- **a**, medium- **b**, and high-magnification **c** FESEM images and TEM images **d** of ZnO NPs [10]. Adapted from Mehar et al. [10].

molecular clusters are formed, which subsequently coalesce into particles. The type of reactant gas as well as the rate of cooling in this process regulate the particle parameters [9]. Low energy cluster beam deposition, a variation of this technique used to create ZnO NPs, is employed when the supersaturated gas is further subjected to supersonic adiabatic expansion.

2.2 *Top-Down Approach*

The most affordable means of synthesizing nanomaterials in large quantities are through mechanical techniques. Ball milling is arguably the most straightforward method. It creates nanomaterials using mechanical attrition where the kinetic energy generated from the grinding media reduces or breaks down the material. An industrial technique known as compaction and consolidation "puts nanomaterials back together" to create materials which have better characteristics. In this method, metallic alloys can be created.

Industry uses a lot of top-down mechanical methods. Here, we focus on thermal methods, which are a broad category where heat is supplied to the fabrication process.

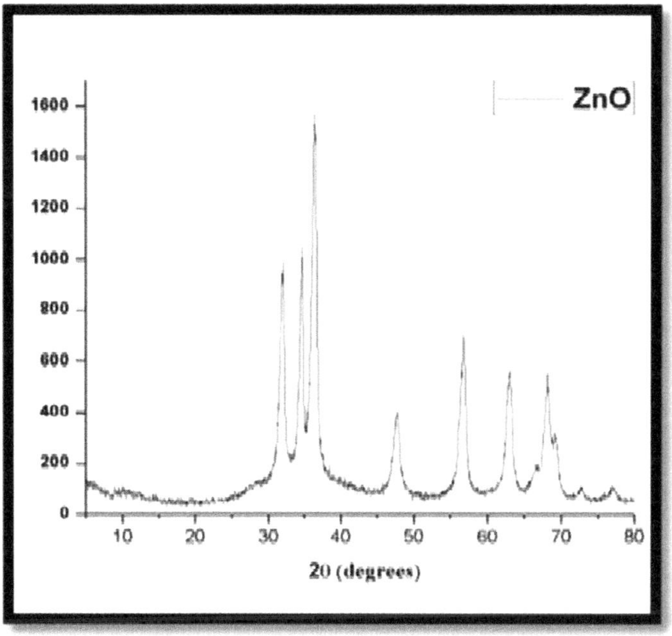

Fig. 5 XRD patterns of ZnO Nanoparticles [10]. [10] Adapted from Mehar et al.

One of these methods for creating nano-thread materials is electrospinning. The methods like solar, electrical, or thermal requires an excessive amount of energy. The first regulated method for creating carbon nanotubes was arc discharge. Laser ablation is also quite promising. Controlling quality and the possibility of upscaling is the issue. Many top-down chemical fabrication techniques, like anodizing, are common industrial processes and are always simple to scale up. It is well known that lithographic techniques, despite being energy-intensive and necessitating costly facilities and equipment, are top-down techniques that primarily produce micron-sized features. For many years, lithography technique has been deployed for printing circuit boards and the fabrication of microprocessors. Future efforts to reduce size will be expensive since they will need stronger sources (such as high-energy electron beams and sources with shorter wavelengths), facilities, and support equipment. Although it isn't lithography by conventional criteria, nanoimprint lithography (NIL). It resembles template synthesis more. To create a pattern, a template material is created first, and it is then stamped into a soft polymeric substance. Both the stamped material and the template stamp are created using a top-down approach. Latex spheres facilitate the creation of a templated matrix in nanosphere lithography. Thus, we may also refer to these methods as template processes. Creating a poreshaped oxide by anodizing aluminium is a straightforward method of creating nanomaterials.

Both the porous structure and any substance generated within are nanomaterials. In many respects, porous membranes constitute the ideal template. There is a fresh wave of nano bottom-up techniques emerging. Numerous innovative techniques are quite cost efficient and result in higher yields. But main drawback is the creation of a long-term order. The new techniques include nanosphere lithography and nanolithography or dip pen method.

2.2.1 Lithography

One kind of top-down approach that is very popular in the semiconductor industry is the lithographic technique. Some common lithographic methods are dip-pen nanolithography, photolithography, focused-ion beam lithography, x-ray lithography, electron-beam lithography. Although these techniques have greatly aided in the development of 1D nanostructures, large-scale 1D nanostructures continue to face difficulties due to their high cost and low throughput. Lithographic techniques are utilized to generate a variety of designed geometry patterns on solid surfaces because of their inherent benefits for the synthesis of 1D nanostructures, including the ability to fabricate large-scale and well-aligned arrays. A schematic representation of the lithography procedures used to create different designed geometry patterns is shown in Fig. 6. The 1D nanostructures can grow in the intended geometry using the lithographic patterns as templates during PVD or particularly glancing angle deposition. 1D nanostructures can form on preferred locations such as shadow sites, V-grooves, cleaved edges, and step edges. The process of photolithography has the following steps:

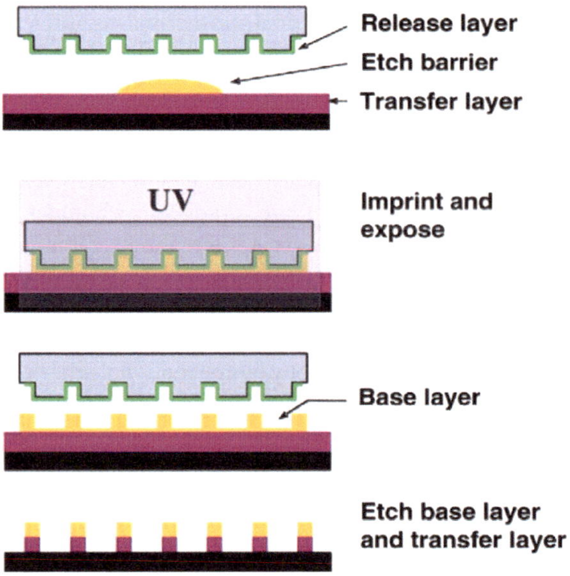

Fig. 6 **a** Schematic representation of the lithographic processes. Adapted from Rasnic D.J. [11]

Coating: A photoresist coating is applied to a substrate, often silicon or another appropriate material. This light-sensitive polymer will be applied to the substrate to produce a pattern.

Exposure: A photomask with the required pattern exposes the coated substrate to ultraviolet (UV) radiation. In the exposed portions, the UV light changes the photoresist chemically.

Development: Depending on whether a positive or negative photoresist was employed, the exposed substrate is subsequently treated with a developer solution that eliminates exposed or unexposed portions of the photoresist.

3 Metal Oxides for Optoelectronic Applications

In real-world applications, there are significant functions which can be performed by metallic oxides (MOs). Because of the strong electronegativity of oxygen, these compounds are stable over the long term when they combine positive metallic and negative oxygen ions. They can be categorised as semiconductors or insulators because of the bandgap that forms in 1–10 eV range when oxygen atoms and metal atoms are combined in the lattice of crystal. Depending on the specific oxide's crystallographic shape, these values might change.

Large-area electronics have seen new developments since the development of hydrogenated amorphous silicon (a-Si H). It is crucial for thin-film transistors (TFTs), which are utilized in display devices. However, its low value of carrier mobility (– 0.5–1.0 cm^2 V^{-1} s^{-1}), lower current carrying capacity, low mechanical flexibility and optical opacity make it unsuitable for these applications. MOs have some special qualities over silicon and other III-V group semiconductors, like optical transparency, tolerance of mechanical stress as well as large carrier mobilities, and affinity towards photoactive and organic dielectric materials due to which scientists are working on these materials. Furthermore, room temperature conditions (~25 °C under air) can be used to produce electronic-grade MO thin films using solution-phase and vapour phase techniques. This broadens their application to high-value products like solar cells, low-cost circuits and flexible organic light-emitting diode (OLED) displays. Metal and MO nanoparticles are essential for improving optoelectronic device performance. They are added to the devices to enhance incident light scattering, which increases the quantity of photons absorbed in the active layer. It has been demonstrated that adding metallic nanoparticles, such as silver (Ag) and gold (Au), improves the optical and structural characteristics of polymer solar cells (PSCs), hence increasing their performance. Furthermore, it has been discovered that adding metal oxide nanoparticles, like CuO and TiO_2, improves the performance of organic-based solar cells. Moreover, customized nanostructured electrode architectures made of metal and MO nanoparticles could improve the conductivity and transparency of thin-film devices like light emitting devices (LEDs) and solar cells.

3.1 MOs as Light-Emitting Devices

Since nanoparticles tend to have large quantum yields of luminescence, they may find use in light-emitting devices. To emit light in the necessary and restricted spectral range, they need to have a band gap. The first effective implementation of the LED concept was achieved in 1994 using CdSe QDs. The emission from CdSe layers occurred at smaller voltages. The emission colour could be changed from red to yellow by varying the size of the nanocrystal. This happens due to the quantum confinement effect and at greater voltages, the green light is emitted from the polymer layer. In this way, voltage tunability of colour is possible for this device. Although this device's effectiveness was only 0.01%, many other devices with a similar structure that were based on NPs have subsequently been developed. ZnO is the only member of the MO group that meets the requirement for a direct band gap, or light emitting device, of more than 3 eV at ambient temperature [12]. In a standard Quantum LED, the light emitting layer of quantum dots like in the form $Cd_{1-x}Zn_xSe$ is placed between the electron transport layer (ETL) made of ZnO nanoparticles and the organic hole transport layer (HTL) made of CBP as given in Fig. 7a. The transparent conducting oxide (TCO) above the ETL is mostly made of ITO (Indium Tin Oxide) and sends the electrons into the ETL. The holes are send from the back contact to the HTL. Previously for LEDs, back contact (mostly Al) was used for injection of electrons and ITO for injection of holes.

After the forward bias current is applied, the holes from HTL and electrons from ETL are send to the QD layer. The ZnO nanocrystals layer (ETL) regulates the optimum charge balance for the QD emitters in addition to facilitation of electron transfer. The QDs are excited by the two simultaneous processes. The one which is called direct charge injection happens when electrons are trapped at the QD layer due to the respective alignment of energy of the lowest unoccupied molecular orbital (LUMO) levels of the ETL, HTL and QDs as shown in the Fig. 7(b). Secondly, after the barrier decreases for injection of the holes from the HTL for the charged QDs, the excitons on the QD layer recombine with the emission of radiation of bright coloured electroluminescence as shown in Fig. 7c. The performance of this QLED could be varied by varying the path of the electroluminescent recombination zone inside the QD film from the interface of QD-ZnO. Molybdenum trioxide (MoO_3) was as the component for hole injection layer decreases the operating voltage and enhances the performance of the device and its lifespan. The surface treatment of the ZnO layer (ETL) with ethanolamine improves the optical performance of this QLED. Further, optimizing the annealing temperature of the ZnO sol–gel electron transport layer increases the efficiency of this QLED by more than three times. This device offers lifespan of hundreds of hours of brightness at 100 cd/m^2 and peak brightness greater than 100,000 cd/m^2.

ZnO being *n*-type shows oxygen vacancies, and points defects leading to high carrier mobility and stability because of which it is a great candidate for use as charge transport layers. Generally, ZnO, TiO_2 and SnO_2 are used as ETL whereas *p*-type metal oxides like MoO_3, NiO_2, V_2O_5, and WO_3 are as HTL in the heterostructured

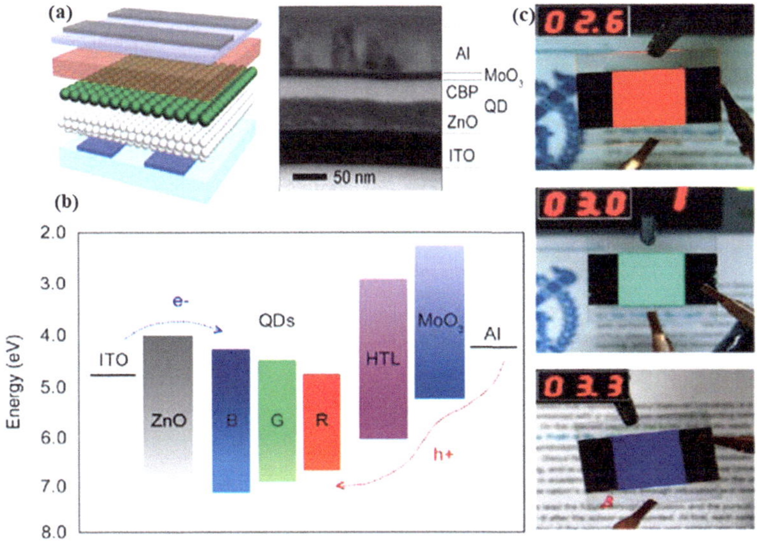

Fig. 7 **a** Schematic diagram and TEM image of the cross section of the inverted Quantum Light Emitting Diode (QLED). **b** Representation of energy level diagram of QLED where colours red(R), green (G) and blue (B) emitted by the QD layer is located between ZnO based electron transport layer (ETL) and CBP based hole transport layer (HTL) [13]. Copyright 2012, American Chemical Society. Kwak et al. [13]

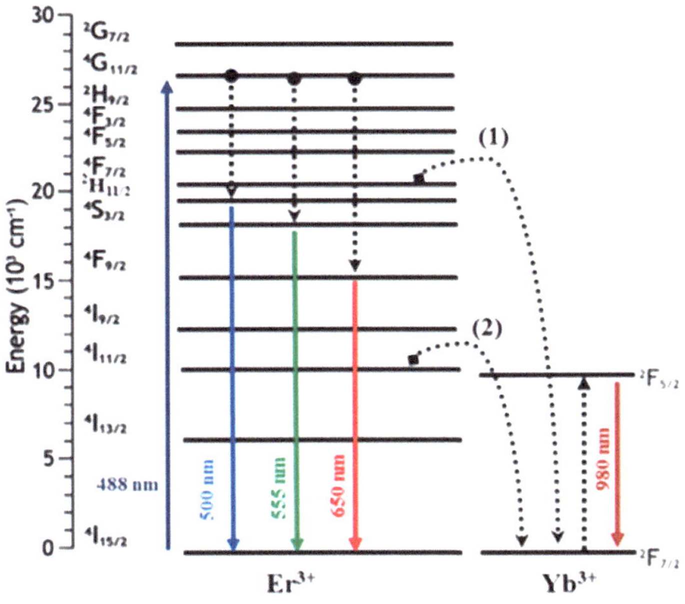

Fig. 8 Energy level diagrams for transitions in Er3+ levels and energy transfer occuring among Er3+/Yb3+ after excitation at 378 nm [16]. Adapted from Derouiche M. [16]

Table 1 List of the different metal oxides with their band gap values

Material	Synthesis route	Band gap (eV)
TiO_2	Sol–gel	4.0
SnO_2	Sol–gel	4.1
MoO_3	Sol–gel	4.9
NiO_2	Sol–gel	4.8
V_2O_5	Evaporation	5.3
WO_3	Evaporation	6.5

LED device. Given below is the table of synthesis methods and energy band levels of some metal oxides used in LEDs. Table 1 shows the different metal oxides with their synthesis route and band gap value.

3.2 MO NPs as Charge Transporting Layers

They are being utilised both in bulk and nanoscale forms to prepare charge-transporting layers (CTL) [14]. This permits the use of significantly greater current densities, and the development of organic-free LED, greatly promising because organic materials tend to decay quickly after being exposed to dampness. Applications for CuO as the hole transporting layer (HTL), ZrO_2 nanoparticles as the electron transporting layer (ETL), QD LED with MoO_3 acting as the HTL and ZnO/TiO_2 as the ETL can all be found in the literature. Solar cells, photodetectors, and other light-emitting devices have all been built using these layers. Particularly in organic dye-sensitized solar cells (DSSC), TiO_2 plays a crucial function as a charge-transporting layer whose surface is utilized for dye absorption. Titania's mesoporosity could be produced through the sintering of TiO_2 NPs, increasing the active surface area thereby greatly improving the solar cell's ability to harvest light.

3.3 MO NPs as Light-Converting Medium

One other application for NPs' photoluminescence effect is the development of a light-converting layer. Since traditional homojunction solar cells cannot utilise the entire solar spectrum like near infrared (NIR) and ultraviolet (UV) region, it may find usage, for example, in photovoltaics. It has been suggested that ZnO NPs function as down-converting (DC) agents, able to absorb UV light and reemit it in the visible spectrum [15]. Doping with rare earth elements could improve their efficiency and luminescence time. The nanoparticles of MOs could also facilitate the up-conversion (UC) mechanism that involves absorbing of the near-infrared photons to produce visible-range photon emission. For this application, the host matrix for REE like Er,

Ho, Nd, and Yb has been the MO NPS of, for example, ZnO, ZrO_2, $BaTiO_3$, and TiO_2.

Figure 8 depicts the emission spectra of Er, Yb: ZnO aerogels while down-converting which is caused by to 4f-4f transition of Er ions in ZnO [16]. At 378 nm excitation, Er, Yb: ZnO shows excellent luminescence both in the infrared and visible ranges. The 2H9/2→4I15/2 transition causes blue emission in the 490-500 nm range. Whereas 4H11/2→4I15/2 (515-535 nm) transition and 4S3/2→4I15/2 (540-565 nm) cause green emission. Similarly, the 4F9/2→4I15/2 transition in the 640-685 nm range causes red emission. The 4I9/2→4I15/2 transition between 790-880 nm leads to infrared emissions while emission at 980 nm is because of the 4I11/2→4I15/2 transition. In the down-converting process, Er3+ are the donors and Yb3+ are the acceptors. For down-conversion, donors must have an energy level greater than 20,000 cm-1 with intermediate energy level of 10,000 cm-1. After the powder is excited by 488 nm wavelength, the relaxation energy from 4G11/2 to 4F7/2 is released and in turn resonant cross relaxation occurs. In cross-relaxation, ions of Er3+ get back to the 4I11/2 level from 4F7/2 leading to the excitation of Yb3+ ions from ground level 2F7/2 to 2F5/2 level as shown in the figure. Similarly, from the intermediate level 4I11/2, Er3+ comes back to ground level 4I15/2 with the release of energy to other Yb3+ ions. This complete phenomenon occurs with the absorption of a single photon (UV range) and the emission of two photons (NIR range). The emissions at 550nm and 670 nm wavelengths confirm that the incident photons get absorbed by the ZnO matrix. ZnO codoped with Er, Yb excites at around 378 nm which helps in calculating the band gap energy of ZnO.

3.4 *Applications as Transparent Conducting Oxides TCO*

Our goal in the transparent conducting oxide (TCO) group is to create and comprehend TCO materials and their applications. We combine basic physical investigations, traditional deposition of thin film, and characterisation approaches to create TCO films which could meet unique utility requirements in solar cells as well as numerous cutting-edge optical devices. Currently, perovskite or silicon with solar cells, Si heterojunction solar cells and OLEDs are all using TCO thin films as electrodes.

The resistivity of the transparent conductor should be no more than 10^{-3} Ω cm, and its average transmittance should be greater than 80% for the visible spectrum. These are the general electro-optical parameters. Consequently, the band gap energy greater than 3.0 eV and minimum carrier concentration value of 1020 cm^3 are necessary for a semiconductor material to be considered for this application. The parameters of these kinds are common to most types of MOs, the most widely researched materials. Amongst the several materials examined, ZnO and its Ga and Al-doped variants, known as GZO and AZO are particularly intriguing and eco-friendly materials [17]. Recent developments in the techniques for depositing these thin films have made it possible to reach resistivity results that are comparable to those of ITO. Making a printing paste with nanoparticles of ITO dissolved in a solvent and applying it to a polymer substrate of polymer is one way to address the drawbacks of the ITO bulk electrode [18]. The accomplishment of a resistance value 0.035 Ω cm along with 89% optical transmittance through UV treatment and annealing in different atmospheres has been demonstrated. These optoelectronic features are satisfactory enough to use the NP layer as the surface contact ifor organically made photodiodes.

4 Conclusions

The general characteristics of MONPs have been discussed along with a list of some of the MONPs used in optoelectronics. It is important to remember that they are merely the very beginning in the field of optoelectronics. MONPs are currently the best option as transparent electrodes. Advanced polymers and carbon nanostructures have some of the same characteristics, achieving somewhat better but still approximate values. However, MO NPs are being thoroughly investigated as a substitute material for some of the rare and expensive semiconductors for the bulk of optoelectronic applications. As a substitute material for some of the scarce and expensive semiconductors, NPs are the subject of much investigation. The progressive transition from Si-centered, all-inorganic devices to flexible, lightweight hybrid organic–inorganic devices is a prominent trend in electronics [19]. Although the MO NPs appear to be a good fit, there are still several barriers to overcome before their practical applications can be commercialized. Either an attempt should be made to improve the prototype devices' efficacy, or emphasis should be placed on reducing the cost of the manufacturing and deposition process. In this way, with their hierarchical structure and nanoarchitecture as well as the optimisation of their properties, the MO nanoparticles could be used in the field of applied nanotechnology enhancing the performance of devices like gas sensors, biotechnology, solar cells, rectifiers batteries, antenna as well as in food, pharmaceuticals, textiles, drones, aerospace and defence applications.

References

1. Chavali, M.S., Nikolova, M.P.: SN Metal oxide nanoparticles and their applications in nanotechnology, Appl. Sci. **1**, 607–610 (2019)
2. Negrescu, A.M., Killian, M.S., Raghu, S.N.V., Schmuki, P., Mazare, A., Cimpean, A.J.: 2.Metal Oxide Nanoparticles: Review of Synthesis, Characterization and Biological Effects. Funct. Biomater. **13**, 274–280 (2022)
3. Shipra, G., Manoj, T.: Cent. 2. 3.A review on the synthesis of TiO2 nanoparticles by solution route. Cent. Eur. J. Chem. **2**, 279–294 (2012)
4. Yoldas, B.E.: Monolithic glass formation by chemical polymerization, J. Mater. Sci. **14**, 1843–1849 (1979)
5. Sahal, M., et al.: Structural, electrical and optical properties of ZnO thin films deposited by sol-gel method. Microelectron. J. **12**, 1425–1428 (2008)

6. Hui, L., et al.: Structure, varistor properties, and electrical stability of ZnO thin films. Mater. Lett. **27**, 2321–2323 (2009)
7. Noh, Y.-J., et al.: Transparent and flexible amorphous InZnAlO films grown by roll-to-roll sputtering for acidic buffer-free flexible organic solar cells. Org. Electron. **24**, 227–233 (2015)
8. Baruwa, A., et al.: Radio Frequency Magnetron Sputtering Coatings of Biomedical Implants Using Nanostructured Titanium Carbide Thin films. J Bio- Tribo- Corros. **7**, 138 (2021)
9. Casey, P.: Nanoparticle technologies and applications. Nanostruct. Control Mater. 1–31 (2006)
10. Maher, S., et al.: Synthesis and Characterization of ZnO Nanoparticles Derived from Biomass (Sisymbrium Irio) and Assessment of Potential Anticancer Activity. ACS Omega **8**(18), 15920–15931 (2023)
11. D. J. Resnick, V. Sreenivasan, C. G. Willson, Step & flash imprint lithography. Materials Today, **8**, 34–42 (2005).
12. Borysiewicz, M.A.: ZnO as a Functional Material, a Review. Cryst. **9**, 505 (2019)
13. Kwak, J., et al.: Bright and Efficient Full-Color Colloidal Quantum Dot Light-Emitting Diodes Using an Inverted Device Structure. Nano Lett. **12**, 5, 2362–2366 (2012).
14. Choi, S., Yong, T., Choi, J.: Towards scalability: progress in metal oxide charge transport layers for large area perovskite solar cells. Inorg. Chem. Front. **11**, 50–70 (2024)
15. Kumar, V., et al.: Rare Earth Doped Zinc Oxide Nanophosphor Powder: A Future Material for Solid State Lighting and Solar Cells. ACS Photon. **4**, 2613–2637 (2017)
16. Derouiche, M. et al., Efficient down conversion ZnO codoped (Er, Yb) nanopowders synthesized via sol-gel process for Si solar cell applications. Journal of Radiation Research and Applied Sciences **16**, 100497 (2023).
17. Jun, M.C., Park, S.U., Koh, J.H.: Comparative studies of Al-doped ZnO and Ga-doped ZnO transparent conducting oxide thin films. Nanoscale Res. Lett. **7**, 639 (2012)
18. Liu, H., Zeng, X., Kong, X., Bian, S., Chen, J.: A simple two-step method to fabricate highly transparent ITO/polymer nanocomposite. Appl. Surf. Sci. **258**, 22, 8564–8569 (2012).
19. Xu, X., Sun, L., Shen, K., Zhang, S.: Organic and hybrid organic-inorganic flexible optoelectronics: Recent advances and perspectives. Synth. Met. **256**, 116137 (2019)

New Materials for New Age: Magneto-Dielectric and Multiferroic Materials

Anil Kumar, Anu Gupta, Rituraj Sharma, Hari Mohan Rai, M. Kamal Warshi, Vikash Mishra, Amodini Mishra, and Pankaj R. Sagdeo

Abstract Magnetodielectric (MD) and multiferroics materials have gained significant attention due to their potential for various technological applications. In multiferroics materials at least two primary ferroic orders, such as ferromagnetism, ferroelectricity, and ferroelasticity coexist within a single-phase compound. However, in MD materials one of the electric properties such as dielectric constant can be tuned with the application of magnetic field. Single-phase multiferroic and magnetodielectric materials are rare at room temperature (RT). Among the promising materials, perovskite oxides, especially rare earth manganite's ($RMnO_3$), orthoferrites ($RFeO_3$), rare earth orthochromites ($RCrO_3$) and $AFeO_3$ (A = Ga, Bi, Al), have

Authors Anil Kumar and Anu Gupta contributed equally.

A. Kumar
Department of Physics, Indian Institute of Technology Kanpur, Kanpur, India
e-mail: anilgarry@gmail.com

A. Gupta
Department of Chemical and Biological Physics, Weizmann Institute of Science, Rehovot, Israel

R. Sharma
Centre for Scientific and Applied Research (CSAR) IPS Academy, Indore, India

H. M. Rai
Department of Physics, Govt. P.G. College Tikamgarh, Tikamgarh, India

Department of Physics, Govt. K. R. G. Post Graduate College, Gwalior 474001, Madhya Pradesh, India

M. Kamal Warshi · P. R. Sagdeo (✉)
Material Research Laboratory, Department of Physics, Indian Institute of Technology Indore, Indore, India
e-mail: prs@iiti.ac.in

V. Mishra
Department of Physics, Manipal Institute of Technology, Manipal Academy of Higher Education, Manipal, Karnataka, India

A. Mishra
Department of Physics, Indian Institute of Technology Delhi, Hauz Khas, New Delhi, India

A. Mishra and V. Pathak (eds.), *Smart Materials for Energy Storage and Biomedical Applications*, Engineering Materials, https://doi.org/10.1007/978-3-031-92584-9_14

been extensively investigated for their MD and multiferroic properties. In these materials, magnetic transition temperature can be tuned to RT by doping and the presence of complex magnetic interactions in these materials can induce ferroelectricity below the magnetic transition temperature through magnetic interactions. Therefore, it might be possible to control the spins orientations by a very small magnetic field, which paves a new way to the design RT spintronic devices. This abstract highlights the new type of multiferroic and magneto-dielectric materials and their role in the future device scalability and multifunctionality in next-generation electronic systems.

Keywords Magnetic interaction · Ferroelectricity · Magnetodielctrics · Transition temperature

1 Introduction

Over the past few decades, astonishing research on silicon devices has enabled the transformation of bulky old desktop computers into sleek smartphones. In the "Silicon Age", the applicability of silicon transistors has become the backbone of the microelectronics industry which has established an era of miniaturization also [1–3]. But this ever-increasing demand of technologically advanced yet miniaturized products can't continue within our existing framework and will slow down the speed of technological advancement in the coming years [1–3]. Therefore, it is now essential to research and develop novel multifunctional materials to maintain and improve the standard of living worldwide. Researchers are also looking for new materials for reading, writing and data exchange processes, as multitasking and powerful microelectronic devices that can store and exchange information faster and easier are in demand. In this context, the new technology known as "spintronics", which enables electric fields to regulate the angular momentum of spins and vice versa, can be used on a large scale [4–7]. For next-generation electronic devices such as highly sensitive DC and AC magnetic field sensors, solid-state transformers, electrically tunable microwave filters and electromagnetic optical actuators, the coexistence of magnetization and electric polarization can lead to an additional degree of degree of freedom [8–11].

At this point, it is worth noting here that the primary energy that we are getting from natural gas, coal, or nuclear power plants is *35–40%* efficiently converted and out we consume *10–15%* of this energy in electronic devices [1–3]. This consumption is expected to increase rapidly in the coming years, which will increase the pressure on our energy production resources. Therefore to improve the global standard of living, new materials or emerging technologies should be developed as an alternative solutions [8–11]. As technology advances, the growing demand for data storage and continued miniaturization signal the impending end of conventional data storage techniques.

Earlier many efforts have been done by scientific community towards developing spin-based electronic devices by using perovskite oxides [2, 3, 8–11]. These materials exhibit complex and interesting properties due to the strong coupling between electric charge, spin, and lattice degrees of freedom [2]. One of the fascinating property of these perovskite oxides is their multiferroic (MF) behavior, which means they have at least two primary ferroic orders (ferromagnetic, ferroelectric, and/or ferroelastic) in a single-phase compound [3]. Regardless of the significant research being carried out in this area, only a few materials exhibit multiferroic orders near room temperature. For example, $BiFeO_3$ possesses a strong ferroelectric polarization coupled with antiferromagnetic ordering. However, in thin film form, a coexistence between ferroelectric and weak ferromagnetism occurs due to emergence of epitaxial strain. To achieve room temperature multifunctional applications, various efforts have been made on synthesizing multiferroic heterostructures [2, 3, 8–11].

The combination of ferromagnetic and ferroelectric thin films in these systems leads to multiferroic and magnetodielectric/magnetoelectric behavior. The interfacial properties can be enhanced by magnetostriction, exchange coupling or exchange bias effects when the ferromagnetic layer is formed on the top surface [2, 3, 8–11]. Moreover, controlling the magnetic state by applying a voltage is continuously explored to overcome the current scenario of energy loss due to the Joule effect. Numerous attempts have been made to utilize magnetostrictive devices in situations where the time-reversal symmetry required to reverse a magnet requires an external magnetic field bias [2, 3, 8–11]. Therefore, multiferroic magnetoelectrics that break both time-reversal and inversion symmetry are expected to be crucial for the development of low-energy spintronics in the future. Further, in these materials, magnetism and ferroelectricity can either occur independently or be strongly coupled, leading to the control of electric polarization by an applied magnetic field or vice versa [2, 3]. Therefore, this chapter highlights the new type of perovskite magneto-electric and multiferroic materials and their critical role in the future of electronics, emphasizing their impact on energy efficiency, device scalability, and multifunctionality in next-generation electronic systems.

2 Multiferroics: A Beautiful but Challenging Multi-polar World

As discussed in the introduction part that to sustain and enhance the global standard of living, there is a pressing need for new multifunctional materials. To address this biggest challenge, modern technology is emerging as a solution by investigating new type of magneto-electric (materials in which electric properties can be tuned by magnetic field and vice versa) and multiferroic materials (materials having more than one ferroic order) [12]. Beyond the potential applications of these materials, the fundamental physics underlying these materials is both rich and fascinating. Conventional ferroelectricity mechanisms in perovskites typically require the B site to have an empty d^0 orbital. This requirement is generally seen as incompatible with ferroic orders, which involve partially filled d-orbitals in transition metals. However, not all theoretical and experimental results align with the view that magnetism and ferroelectricity are mutually exclusive.

The mutual coupling of magnetic and electric fields was firstly predicted by the Maxwell's equations. It is well known that changing magnetic field induces an electric field, and moving charges, or electric currents, produce a magnetic field. Additionally, the formal equivalence between the equations governing electrostatics and magnetostatics in polarizable media highlights similarities between ferromagnetism and ferroelectricity, such as their hysteresis behaviour in response to external fields and the anomalies (thermal, structural, mechanical, dielectric, and magnetic) observed at critical temperatures [2, 3, 12]. These coupling phenomena and similarities between electric dipoles and spins support the potential for combining ferroelectricity and magnetism within single-phase materials, suggesting the possibility of magnetoelectric (ME) effects. Further, achieving simultaneous ferroelectric and ferromagnetic orders well above room temperature is a one of biggest challenge in front of scientific community [2]. Although certain materials exhibit both ferroelectricity and ferromagnetism, however the coupling between these properties at room temperature is often very less.

3 Recent Research Work Done

In multiferroics, ferroelectricity and ferromagnetism can occur independently or be strongly coupled. In particular, ferroelectricity can be triggered by the canting of magnetic spins or spin chirality. In contrast to traditional ferroelectricity as observed in $BaTiO_3$, ferroelectric behavior in multiferroic materials is not due to the off-center displacement of small transition metal (TM) cations [2]. Instead, the d orbitals of the TM cation must be formally unoccupied. The first experiment to directly measure the interaction between electrical and magnetic parameters was carried out by Wilhelm Röntgen in 1888. The authors proved that in a moving body, both the dielectric displacement D and the magnetic induction B are influenced by electric and magnetic fields [2]. In 1894, Pierre Curie proposed theoretically that an electric or magnetic field can cause magnetizationor polarization in certain materials due to symmetry arguments [2]. This idea was largely overlooked until the 1950s, when Lev Landau and Evgeny Lifshitz briefly discussed magnetoelectric terminology in their book "Electrodynamics of Continuous Media [13]". In the 1960s, Igor Dzyaloshinskii ushered in a "golden age" of magnetoelectric research [14] with his phenomenological theory predicting the magnetoelectric effect in chromium oxide, which was experimentally confirmed later that year by D. N. Astrov [14].

Ongoing theoretical and experimental research is investigating ways of circumventing the above-mentioned rule of empty *d* orbitals. One approach is to use epitaxial stretches to influence the balance between covalent bonding and Coulomb repulsion, thereby promoting off-centering even in cations with partially filled d orbitals.

Recently, significant research has been carried out on these perovskite materials by various research groups [2, 3, 12, 15, 16]. Recently, Wang et al. reported the multiferroicity in the epitaxial films of $BiFeO_3$ [17]. The ferroelectric hysteresis loop showed an out-of-plane saturation polarization of 60 $\mu C/cm^2$, which corresponded to ~ 90 $\mu C/cm^2$ along the [111] direction, the highest known at that time. Nevertheless, the Fe ions align antiferromagnetically due to strong super-exchange interactions, although there is a slight tilt leading to weak ferromagnetism.

Acharya et al. [15] introduced a novel multiferroic compound, $LaFeO_3$, which exhibited coexistence of coupled ferroelectric and magnetic orderings [15]. The sample shows both spontaneous electric polarization and magnetization, along with reasonable magnetocapacitance percentage. Due to the spontaneous electric/magnetic polarization of the sample, the observed loop clearly shows the ferroelectric/ferromagnetic arrangement of the electric/magnetic dipole moments. This shows that the present sample exhibits multiferroic activity or both electric and magnetic orderings. When an applied magnetic field initiates magnetic ordering in MF materials, the dielectric constant is affected. The dielectric constant of the sample was tested both with and without the presence of a magnetic field to detect this phenomenon. The figure (see Fig. 4 in Ref. [15]) clearly shows that there is a positive coupling as the dielectric constant (ε') increases with the applied magnetic field (H). For H at 1.3 T [15], the maximum magnetocapacitance (MC) is about 5%. In addition, the origin of the coupling was clarified in terms of the strain experienced by the sample while it is exposed to a magnetic field. This strain arises from the interaction between the magnetic and ferroelectric domains, which in turn induces a voltage that generates an electric field within the ferroelectric domain. This strain caused by the magnetic field is discussed in more detail in the following sections.

In the recent study, Rai et al. [11] investigated the control of electric parameters, specifically the dielectric constant, in Mn-doped $LaGaO_3$ samples using a magnetic field. Further, it has been observed by the authors is that the observed coupling between electric and magnetic parameters is found to be intrinsic over the entire range of probing frequencies as shown in Fig. 1. Further possible mechanism for this coupling has been explained as follows; on applying a 0.4 T magnetic field aligns the magnetic moments of Mn ions, which can change their orbital arrangement due to the interconnected nature of charge, spin, and orbital properties [11]. This realignment can modify the orthorhombic strain and thus affect the dielectric constant (ε'). The change in ε' at 0.4 T is due to magnetostriction, which alters the lattice parameters and bond distances/angles.

Further, Warshi et al. [18] reported the magnetodielectric properties of well $GaFeO_3$ sample for its multiferroic and magnetoelectric (MD) properties as depicted in Fig. 2. Even at ambient temperature, the dielectric constant varies significantly when a magnetic field is applied in the MHz frequency range, as shown in Fig. 2. Around 100 kHz, MD coupling is visible in this oxide, with $GaFeO_3$ exhibiting a magnetocapacitance of about 25% at 2 T. This significant magnetocapacitance is notable and aligns with earlier published data [18, 19].

The origins of natural and magnetic field-induced polar ordering in Cr-doped rare earth perovskite samples were recently investigated by Kumar et al. [20] and

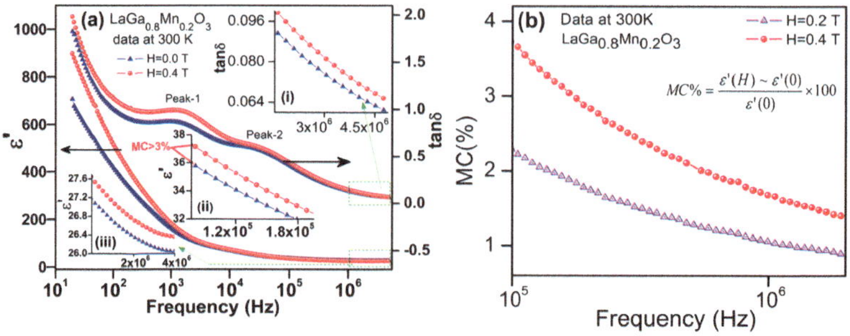

Fig. 1 **a** Dielectric loss (tan δ) and room-temperature dielectric constant (ε′) for $LaGa_{0.80}Mn_{0.20}O_3$ as a function of frequency and magnetic field. **b** Variation of MC% as a function of frequencies. Data reproduced from Ref. [11]

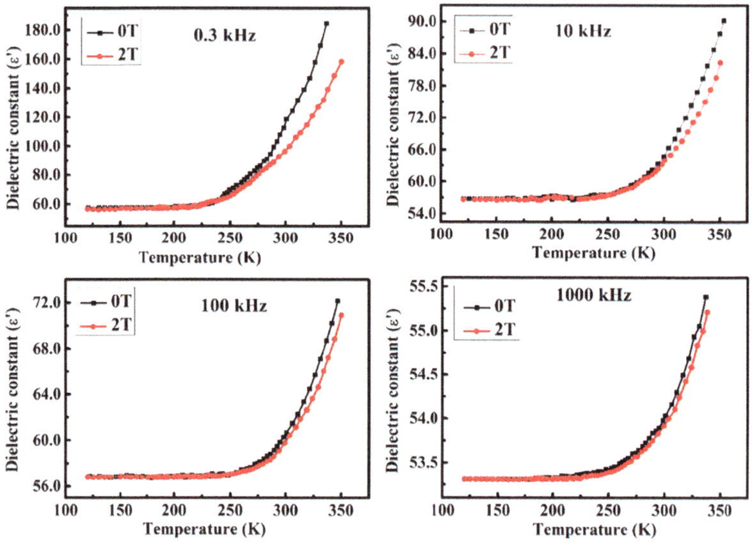

Fig. 2 The measured magneto-dielectric effect of the $GaFeO_3$ sample at different frequencies. Data reproduced from Ref. [18]

the corresponding results are shown in Fig. 3. It was discovered that the magnetic transition temperature in the $PrFeO_3$ sample can be tuned to room temperature by doping the Fe site with Cr. This is due to the close correlation between the orbital, lattice and spin degrees of freedom in these materials. Therefore, it is possible to tune spin orientations with a minimum magnetic field via magnetostriction, which improves the feasibility of magneto-dielectric coupling and offers prospects for the development of room temperature devices [20]. Temperature-dependent dielectric (TD) measurements at 1 MHz, as shown in Fig. 3a, reveal the presence of magneto-dielectric coupling. An anomaly in the permittivity data around the magnetic ordering temperature of ~ 280 K, confirmed by the first derivative of the ε versus T curve (inset of Fig. 3a), appears as a cusp near the magnetic transition temperature, indicating a magnetodielectric effect near room temperature. Similar peak characteristics in $GaFeO_3$ and $YMnO_3$ are associated with magnetodielectric effects [20]. This behaviour is attributed to antiferromagnetic ordering due to super-exchange interactions between Fe^{3+}/Cr^{3+} ions. Above 280 K, the dielectric constant varies normally, but below this temperature, internal hyperfine magnetic field interactions change the trend. The fundamental origin of magneto-dielectric coupling is understood by magnetic field-dependent Raman measurements.

The symmetric stretching of the Fe/Cr–O bonds is represented by the Raman mode at about 690 cm^{-1}, while the octahedral tilt is associated with the softer Ag modes at lower wavenumbers [20]. Magneto-expansion effects are indicated

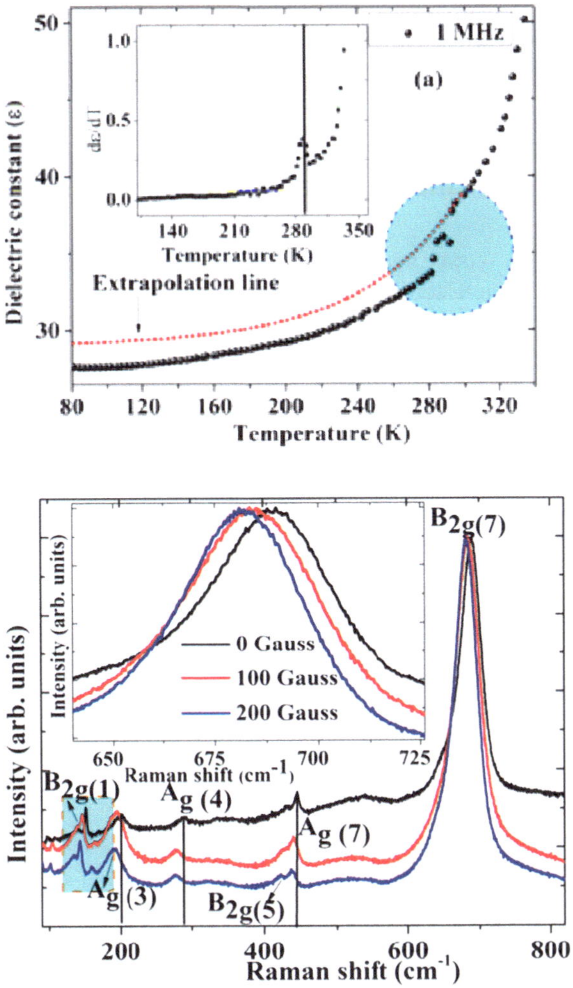

Fig. 3 The temperature dependence of the dielectric constant at 1 MHz is shown in (**a**). The first derivative of the dielectric constant as a function of temperature is shown in the inset. The magnetic field-dependent Raman spectra of the Cr-doped $PrFeO_3$ sample are shown in (**b**). Data reproduced from Ref. [20]

by the magnetic field-dependent Raman spectra in Fig. 3b, which show considerable sensitivity to applied magnetic fields. Moreover, the rotation of spin-coupled Fe orbitals under the influence of a magnetic field explains the emergence of SPC-induced magneto-dielectric coupling [20]. The alignment or rotation of magnetic moments within sublattices facilitates the realignment of orbitals, which can change the orthorhombic tension or cell volume. Similar coupling has been observed in materials such as $TbMnO_3$, $Ni_3V_2O_8$, $MnWO_4$, and $BaNiF_4$, which is attributed to the reverse Dzyaloshinski–Moriya (DM) effect [2, 16, 21–23]. This effect manifests itself in systems with non-collinear spin structures, often in the form of spiral magnetic configurations where the spins rotate about an axis perpendicular to the propagation vector of the spiral, requiring direct involvement of spin–orbit interaction. It is also worth noting that magneto-dielectric (MD) and magneto-electric (ME) effects typically require high magnetic fields, often several Tesla, and temperatures below room temperature, which limits their practical applicability for low-field and room-temperature applications. In addition, the presence of room temperature coupling in these materials significantly increases their technological importance.

4 Conclusion

A new type of materials in which electric/magnetic properties can be controlled with magnetic/electric field at room temperature are essential for advancing technology due to their unique physical properties and their practical applicability, energy efficiency, and multifunctionality. The results presented in this book chapter reveals that intrinsic MD coupling can be understood in terms of strain induced in the sample in the presence of an externally applied magnetic field, which paves a way to the design of RT-based devices. Further the research on multiferroic and MD materials may enable the development of innovative and cost-effective devices that can operate efficiently under standard conditions. For future high-density recording media, these materials could also represent an alternative to heat-assisted magnetic recording. As a final remark, the presence of coupled electric and magnetic interactions at room temperature boosts their commercial and practical appeal to revolutionize modern electronic systems and drive future technological advancements.

Acknowledgements Author Anil Kumar acknowledges Indian Institute of Technology Kanpur and Department of Science and Technology for financial support under grant No. (DST/INSPIRE/IFF/BATCH- 20/2024-2025/IFA 23-PH 307). The authors acknowledge DST-FIST (Grant No. SR/FST/PSI-225/2016) for providing funding for the Raman spectrometer.

References

1. Aleksiyko, R., Berkowski, M., Fink-Finowicki, J., Byszewski, P., Diduszko, R., Kowalska, E. (eds.): A. Rogalski, K. Adamiec and P. Madejczyk, Zakopane, Poland, pp. 50–54 (2001)
2. Spaldin, N.: MRS Bull. **42**, 385–390 (2017)
3. Eerenstein, W., Mathur, N.D., Scott, J.F.: Nature **442**, 759–765 (2006)
4. Li, X., Tang, C., Ai, M., Dong, L., Xu, Z.: Chem. Mater. **22**, 4879–4889 (2010)
5. Kavanagh, C.M., Goff, R.J., Daoud-Aladine, A., Lightfoot, P., Morrison, F.D.: Chem. Mater. **24**, 4563–4571 (2012)
6. Sivakumar, M., Gedanken, A., Bhattacharya, D., Brukental, I., Yeshurun, Y., Zhong, W., Du, Y.W., Felner, I., Nowik, I.: Chem. Mater. **16**, 3623–3632 (2004)
7. Vaqueiro, P., López-Quintela, M.A.: Chem. Mater. **9**, 2836–2841 (1997)
8. Krohns, S., Lunkenheimer, P., Meissner, S., Reller, A., Gleich, B., Rathgeber, A., Gaugler, T., Buhl, H.U., Sinclair, D.C., Loidl, A.: Nat. Mater. **10**, nmat3180 (2011)
9. Sagdeo, A., Gautam, K., Sagdeo, P.R., Singh, M.N., Gupta, S.M., Nigam, A.K., Rawat, R., Sinha, A.K., Ghosh, H., Ganguli, T., Chakrabarti, A.: Appl. Phys. Lett. **105**, 042906 (2014)
10. Sagdeo, P.R., Anwar, S., Lalla, N.P.: Phys. Rev. B **74**, 214118 (2006)
11. Mohan Rai, H., Saxena, S.K., Mishra, V., Sagdeo, A., Rajput, P., Kumar, R., Sagdeo, P.R.: J. Mater. Chem. C **4**, 10876–10886 (2016)
12. Khomskii, D.: Physics (2009). https://doi.org/10.1103/Physics.2.20
13. Landau, L.D., Lifshitz, E.M.: Electrodynamics of Continuous Media, Internet Archive (1960)
14. Dzyaloshinskii, I.E.: On the magneto-electrical effects in antiferromagnets. Soviet Physics JETP **10**, 628–629 (1960)
15. Acharya, S., Mondal, J., Ghosh, S., Roy, S.K., Chakrabarti, P.K.: Mater. Lett. **64**, 415–418 (2010)
16. Hu, N., Lu, C., Xia, Z., Xiong, R., Fang, P., Shi, J., Liu, J.-M., Appl, A.C.S.: Mater. Interfaces **7**, 26603–26607 (2015)
17. Wang, J., Neaton, J., Zheng, H., Valanoor, N., Ogale, S., Liu, B., Viehland, D., Vaithyanathan, V., Schlom, D., Waghmare, U., Spaldin, N., Rabe, K., Wuttig, M., Ramesh, R.: Science **299**, 1719–1722 (2003)
18. Warshi, M.K., Kumar, A., Sagdeo, P.R.: Meas. Sci. Technol. **30**, 125901 (2019)
19. Shireen, A., Saha, R., Mandal, P., Sundaresan, A., Rao, C.N.R.: J. Mater. Chem. **21**, 57–59 (2011)
20. Kumar, A., Warshi, M.K., Sagdeo, A., Krzystyniak, M., Rudiifmmode, S., Adroja, D.T., da Silva, I., Sagdeo, P.R.: Phys. Rev. B **104**, 035101 (2021)
21. Fiebig, M., Lottermoser, T., Meier, D., Trassin, M.: Nat. Rev. Mater. **1**, 16046 (2016)
22. Harris, A.B., Yildirim, T., Aharony, A., Entin-Wohlman, O.: Phys. Rev. B **73**, 184433 (2006)
23. Wang, C., Guo, G.-C., He, L.: Phys. Rev. Lett. **99**, 177202 (2007)

GPSR Compliance

The European Union's (EU) General Product Safety Regulation (GPSR) is a set of rules that requires consumer products to be safe and our obligations to ensure this.

If you have any concerns about our products, you can contact us on ProductSafety@springernature.com

In case Publisher is established outside the EU, the EU authorized representative is:

Springer Nature Customer Service Center GmbH
Europaplatz 3
69115 Heidelberg, Germany

Batch number: 10755216

Printed by Printforce, the Netherlands